unesco

European Commission

IWA PUBLISHING

Implementing the Water–Energy–Food–Ecosystems Nexus and Achieving the Sustainable Development Goals

Edited by

C. Carmona-Moreno, E. Crestaz, Y. Cimmarrusti, F. Farinosi, M. Biedler, A. Amani, A. Mishra and A. Carmona-Gutierrez

Published in 2021 by the United Nations Educational, Scientific and Cultural Organization (UNESCO), 7, place de Fontenoy, 75007 Paris, France; European Commission, Joint Research Centre, Institute on Sustainable Resources, Water and Marine Resources Unit, TP. 121 Via Fermi, 2749 I-21027 ISPRA (VA), Italy and IWA Publishing, the Export Building, Republic, 2 Clove Crescent, East India, London E14 2BE, United Kingdom.

ISBN 978-92-3100473-5 (UNESCO)

Cover photos (from left to right): ©Alex/AdobeStock ©electriceye/AdobeStock ©JillWellington/Pixabay ©Fesus Robert/Shutterstock
Graphic design: PaulSmithDesign
Cover design: PaulSmithDesign
Typeset: Nova Techset
Printed by: Rotolito S.p.A.

SHORT SUMMARY

Addressing the Water–Energy–Food–Ecosystems Nexus to achieve the SDGs

Natural resources are increasingly at risk due to climate change, demographic pressure and economic growth, political instability and forced migration. Globalisation places additional strain on the resources, biodiversity and ecosystems and, through them, on the economies and well-being of populations of affected and neighbouring countries. While approximately 2.2 billion people around the world still lack safely managed drinking water, 789 million people lacked electricity in 2018 and an estimated 25.9 percent of the global population – 2 billion people – were affected by moderate or severe food insecurity in 2019.

It was in this context of water, energy and food constraints that, in 2018, the European Commission's Joint Research Centre (JRC) in partnership with UNESCO's Intergovernmental Hydrological Programme (IHP) launched the project "Water–Energy–Food–Ecosystems (WEFE) Nexus: Analysing solutions for security supply". The WEFE Nexus aims to increase water, energy, food security without compromising ecosystems services. Its components are present in **14** of the **17 SDGs** and are therefore highly relevant in terms of working towards their implementation.

WEFE Nexus components are present in **14** of the **17** SDGs

This publication compiles a number of case studies with the aim of highlighting the importance and benefits of the **WEFE Nexus approach for development cooperation**. It identifies pathways for a more integrated and sustainable use of resources that goes beyond traditional sector-specific development silos.

unesco

"Since wars begin in the minds of men and women it is in the minds of men and women that the defences of peace must be constructed"

European Commission

Contributors

Volume edited by

European Commission

unesco
Intergovernmental Hydrological Programme

With the participation of:

IHE Delft
https://www.un-ihe.org/media

Contents

Foreword

Natural resources are increasingly at risk due to climate change, demographic pressure and economic growth, political instability, and forced migration. Globalization places additional strain on the resources, biodiversity, and ecosystems and, through them, on the economies and well-being of populations of both affected and neighbouring countries. Among the greatest challenges to economic and social development and cooperation are the identification of appropriate and timely adaptation measures in this continuously changing environment. Equally important are the multi-sectorial interlinkages towards achieving the sustainable development goals (SDGs), Paris Agreement targets, the Sendai Framework for Disaster Risk Reduction, the EU Green Deal and emerging challenges such as the coronavirus disease (COVID) crisis.

In 2018, the European Commission's Joint Research Centre (JRC) in partnership with UNESCO-Intergovernmental Hydrological Programme (IHP) launched the project 'Water-Energy-Food-Ecosystems (WEFE) Nexus: Analysing solutions for security supply', supported by the EU's NEXUS Dialogue Programme and co-funded by the European Union and the German Federal Ministry for Economic Cooperation and Development. It addresses efficient implementation of sustainable growth measures by contributing to EU policy objectives and performing analyses of the resilience of water, food, and energy security in societies.

On 25 and 26 January 2018, WEFE experts from universities, international organizations and the European Commission came together to share experiences and ideas in a dialogue on the WEFE Nexus methodology and its support to the SDGs' implementation. Lessons learned, challenges and solutions for applying a Nexus approach to projects and programming were drawn from pilot studies from the regions of Africa, Latin America, South-East Asia, Central Asia, the Middle East, and North Africa. Studies ranged from national to regional and transboundary levels, incorporating both rural and urban contexts. The meeting was organised around three themes: (1) Sustainable Technological Approaches and Solutions; (2) Nexus tools/models/data; and (3) Governance, Finance, Institutions and Cooperation Frameworks for the Nexus. The SDGs were addressed in case studies across the three themes and were part of the Nexus discussions. From 2016 to 2020, the EU-funded ACEWATER II Project, including Human Capacity Development project coordinated by UNESCO-IHP with the AUDA-NEPAD Network of Water Centres of Excellence, implemented Nexus activities in different regions in Africa. Other EU projects in parallell addressed Nexus themes in the Mekrou (Benin, Burkina Faso, and Niger), and the Senegal (Senegal, Mauritania, Mali, and Guinea) transboundary river basins; all of which contributed to a growing knowledge and a better understanding of challenges and opportunities in WEFE Nexus concepts and operationalization.

These case studies provided examples of how hydro-climate, energy and agriculture modelling integrating information from various data sources, including ground-based data, can lead to the development of tools and Spatial Decision Support Systems relevant to achieving water, food and energy security, and help identify and implement best practices in WEFE Nexus assessment. They commonly emphasized the need for cooperation among stakeholders and decision-makers from various sectors, and that existing modelling approaches presented challenges, with data access and sharing, collection and collaborative modelling. Qualitative analysis was also examined as a potential approach to addressing multiple sectors, stimulating a wider Nexus dialogue.

This publication highlights the key outcomes of these exchanges and case studies, aiming to launch a comprehensive guide on operationalizing a WEFE Nexus approach, making it relevant to the key policy areas outlined in the New European Consensus on Development and the European Green Deal. It identifies pathways for the WEFE Nexus approach and a more integrated and sustainable use of resources that goes beyond traditional sector-specific development silos and allows for application at different scales. This is particularly relevant in an evermore globalized and interconnected world, where consultation and collaboration become increasingly important for sustainable development, with a broader institutional and social participation leading to a stronger ownership by partners, stakeholders, and beneficiaries.

Mr. Stephen QUEST
Director General
Joint Research Centre
European Commission

Ms. Shamila NAIR-BEDOUELLE
Assistant Director-General for
Natural Sciences
UNESCO

List of Acronyms

ACEWATER	African networks of Centres of Excellence on WATER Sciences
AFD	Agence Française de Dévelopment
AfDB	African Development Bank
AGCE	Africa Geothermal Center of Excellence
AMCOW	African Ministers' Council on Water
ARC2	African Rainfall Climatology version 2
AU	African Union
AU-IBAR	African Union – Interafrican Bureau for Animal Resources
AWM	Agricultural Water Management
BCM	Billion Cubic meters
BL	BaseLine
BNB	Blue Nile Basin
CAADP	Comprehensive Africa Agriculture Development Programme
CAPP	Central African Power Pool
CDF	Cumulative Distribution Function
CHIRPS	Climate Hazards Group InfraRed Precipitation with Station data
CORDEX	Coordinated Regional Climate Downscaling Experiment
CV	Climate Variability
DG-DEVCO	Directorate-General-for international DEVelopment and Cooperation (now INTPA)
DSS	Decision Support System
EAPP	Eastern African Power Pool
EARS	East African Rift System
EC	European Commission
ECOWAS (CEDEAO)	Economic Community of West African States
EGS	Ecosystems Goods and Services
EL	Evaporation Losses
ENSO	El Niño Southern Oscillation
EPC	Engineering Procurement & Construction
ESCOM	Electricity Supply Corporation of Malawi (Ltd)

ETS	Equitable Threat Score
EU	European Union
FAO	Food and Agriculture Organization
FAR	False Alarm Ratio
FREEWAT	FREE and open-source software tools for WATer resource management
GDC	Geothermal Development Company (of Kenya)
GDP	Gross Domestic Product
GERD	Great Ethiopian Renaissance Dam
GHG	GreenHouse Gas
GIZ	Deutsche Gesellschaft für Internationale Zusammenarbeit
GIS	Geographical Information System
GOF	Goodness Of Fit
GRMF	Geothermal Risk Mitigation Facility (for Eastern Africa)
GW	GroundWater
HCD	Human Capacity Development
HDI	Human Development Index
HEC	Hydrological Engineering Center from the US Army Corps of Engineers
HMS	Hydrologic Modelling System
ICT	Information and Communication Technology
IEA	International Energy Authority
IGA	International Geothermal Association
IGRAC	International Groundwater Resources Assessment Centre
INTPA	Directorate-General-for INTernational PArtnership
IPPs	Independent Power Producers
IRENA	International Renewable Energy Agency
ITCZ	InterTropical Convergence Zone
IWMI	International Water Management Institute
IWRM	Integrated Water Resources Management
JICA	Japan International Cooperation Agency
JRC	Joint Research Centre of the European Commission
KenGen	Kenya Electricity Generating Company
LAC	Latin America and the Caribbean
LCOE	Levelized Cost Of Electricity
LDCs	Least Developed Countries
MBE	Mean Bias Error
MW	MegaWatt
NA	North Africa
NAPP	North African Power Pool
NDF	Nordic Development Fund
NDVI	Normalized Difference Vegetation Index
NRB	Niger River Basin
NSE	Nosch-Sutcliff Efficiency
OECD	Organization for Economic Co-operation and Development
O&G	Oil and Gas
OMVS	Organization pour la Mise en Valeur du fleuve Sénégal
ORC	Organic Rankine Cycle
OUM	Overall Unified Metric
PERSIANN-CDR	Precipitation Estimation from Remotely Sensed Information Using Artificial Neural Networks–Climate Data Record
POD	Probability of Detection

PV	PhotoVoltaic
REC	Regional Economic Community
REV	Risk of Environmental flow Violation
RMSE	Root Mean Squared Error
RSS	Risk of irrigation Supply Shortage
SAPP	Southern African Power Pool
SADC	Southern African Development Community
SARDC	Southern African Research and Documentation Centre
SDGs	Sustainable Development Goals
SEA	Strategic Environmental Assessment
SIDS	Small Island Developing States
SIS	Smallholder Irrigation Schemes
SPI	Standardized Precipitation Index
SRB	Senegal River Basin
SUTRA	Saturated–Unsaturated, variable-density ground-water flow with solute or energy TRAnsport
SWAT	Soil and Water Assessment Tool
T	Temperature
UKZN	University of KwaZulu-Natal
UM	Unified Metric
UN	United Nations
UNDP	United Nations Development Programme
UNEP	United Nations Environmental Programme
UNESCO	United Nations Educational, Scientific and Cultural Organization
UNFCCC	United Nations Framework Convention on Climate Change
USAID	United States Agency for International Development
WASH	Water, Sanitation and Hygiene
WB	World Bank
WEAP	Water and Evaluation And Planning system
WEFE	Water-Energy-Food-Ecosystem Nexus
WEP	Water-Energy Productivity
WHO	World Health Organization
WRC	Water Research Center (Univeristy of Khartoum, Sudan)
WWW	World Wide Web
ZAMCOM	ZAMbezi water course COMmission
ZPC	Zimbabwe Power Company
ZRB	Zambezi River Basin

Background, Purpose, and Target Audience

The Nexus concept has become widely used in the international development community in recent years, not least since the Bonn 2011 Nexus Conference. While the Millennium Development Goals did not include a goal on energy and had no focus on the WEFE Nexus, the global development policy context has significantly changed with the 2030 Agenda and its Sustainable Development Goals (SDGs, 2015), the Paris Agreement on Climate Change (2015), the Addis Ababa Action Agenda on Financing for Development (2015), and the New Urban Agenda (2016) that underline the importance of policy coherence and integrated approaches across traditional development sectors. Last but not least, the **new Green Deal clearly** establishes the link between multiple sectors as already outlined in the New European Consensus on Development.

Current EU development policy provides a general framework and background that justifies a Nexus approach and methodology. Consequently, the New European Consensus on Development: 'Our World, our Dignity, our Future' (June 2017) emphasizes an integrated approach to development and strongly supports the 2030 Agenda with its 17 Sustainable Development Goals (SDGs) that were adopted in September 2015 as a global framework for sustainable development action. The 2030 Agenda itself emphasizes an integrated approach that can be facilitated by a Nexus methodology. As stated by the former INTPA Director General Stefano MANSERVISI in an interview with DEVEX on 14 June 2018: '**The SDGs are obliging us to work in an integrated way and not in silos**'.

The Nexus approach, in its evolution, can now be defined in a number of ways. DG INTPA has supported work on the Nexus from the outset, originally addressing the energy–water–food security (WEF) Nexus. However, ecosystems are increasingly recognized as important contributions to sustainable development solutions and, as a result, the WEF Nexus has evolved into the water–energy–hay–ecosystems Nexus ('WEFE nexus'). In the following sections, the terms 'WEFE Nexus' or 'Nexus' in short refers to this concept unless otherwise stated.

This document brings together a number of contributions, case studies, and experiences with the aim of highlighting the importance and benefits of the WEFE Nexus as an approach and methodology for development cooperation in general. It makes special reference to the main conclusions of the WEFE Nexus specific workshop held in Brussels jointly organized by JRC and INTPA and held on 25–26 January 2018. The workshop was oriented towards implementing and operationalizing the Nexus approach with the objective of improving the sustainability of the intervention projects and programmes based on the experience of the Nexus experts in a variety of projects and regions. This document should be read as a practical guide to fully understand key aspects of WEFE Nexus and is presented in non-technical language for ease of access and uptake for a wider audience.

The initial target audiences for this document are: (i) DG INTPA and other European Commission Directorates-General staff involved in water, energy, food security development cooperation and the environment; (ii) EU Delegations in Partner countries in Africa, Asia, Latin America, and the EU neighbourhood (MENA) region working in these areas; (iii) Development Cooperation Agencies of the EU Member States; and (iv) key development actors including International Organizations such as UN Agencies and NGOs. However, the style of this document also encourages a larger target audience of multiple disciplines in academia and the wider public to access concepts on the Nexus approach.

1

The WEFE Nexus in brief

1.1 THE NEXUS DEFINED

The water–energy–food–ecosystems (WEFE) Nexus is an approach that moves away from the traditional focus on separate entities but rather integrates management and governance across the multiple sectors of food, energy, water, and ecosystems as being complex and inextricably entwine. Globally as well as locally, there is a growing acknowledgment of the interconnectedness between water, energy, food security, and ecosystems. We are already aware that direct inputs of water are needed in the production of food and energy, while energy is required for the storage and distribution of food as well as in water extraction, conveyance, and treatment. Finally, natural resources, ecosystems and their services also underpin water, food, and energy security. Any limitation in inputs from one of these components disturbs the quantity, quality and access of the others to a good Nexus balance. Applying the WEFE Nexus approach helps to improve understanding of the interdependencies across these sectors[1] with a view to fostering integrated solutions in those fields that can facilitate the achievement of development goals such as the sustainable development goals (SDGs).

This approach is optimally coordinated by a central or solid and binding coordination mechanism (allowing for Nexus policy dialogues) where key stakeholders can better identify and prioritize solutions together, benefitting from an overall Nexus perspective. Under the Nexus paradigm, line ministries and key sector actors are guided to consider and integrate priorities of other sector mandates and actors. This can require compromises or accepting decisions that may not *initially* be seen as optimal

[1]**Ecosystem services** are the many and varied benefits to humans provided by the natural environment and from healthy ecosystems. Such ecosystems include, for example, agroecosystems, forest ecosystems, grassland ecosystems, and aquatic ecosystems. These ecosystems, functioning in healthy relationship, offer things such as natural pollination of crops, clean air, extreme weather mitigation, and human mental and physical well-being. Collectively, these benefits are becoming known as 'ecosystem services', and are often integral to the provisioning of clean drinking water, the decomposition of wastes, and resilience and productivity of food ecosystems.

for a single sector, but which a Nexus perspective provides the knowledge and decision-sharing framework to identify and view trade-offs as being strategically beneficial for all sectors involved. In these dialogue and negotiation processes, the direct involvement of the scientific–technical dimension as a supporting element, for providing evidence for informed Nexus thinking and ultimately implementation is essential.

1.2 THE FUNDAMENTAL PRINCIPLES

On an operational level, a complete holistic approach which recognizes that everything is influenced by everything else is intricate and difficult to implement. Choosing a limited number of sectors to work with, such as the WEFE Nexus presents a more workable scale and the links are already intuitively recognized between these specific sectors. Nevertheless, planning for up to four components is far more complex than single sector planning. Figure 1.1 provides a graphic presentation of the Nexus complexity designed by the Food and Agriculture Organization (FAO), showing how it can be simplified by breaking it down into recognizable components of stakeholders, drivers, objectives, and resources.

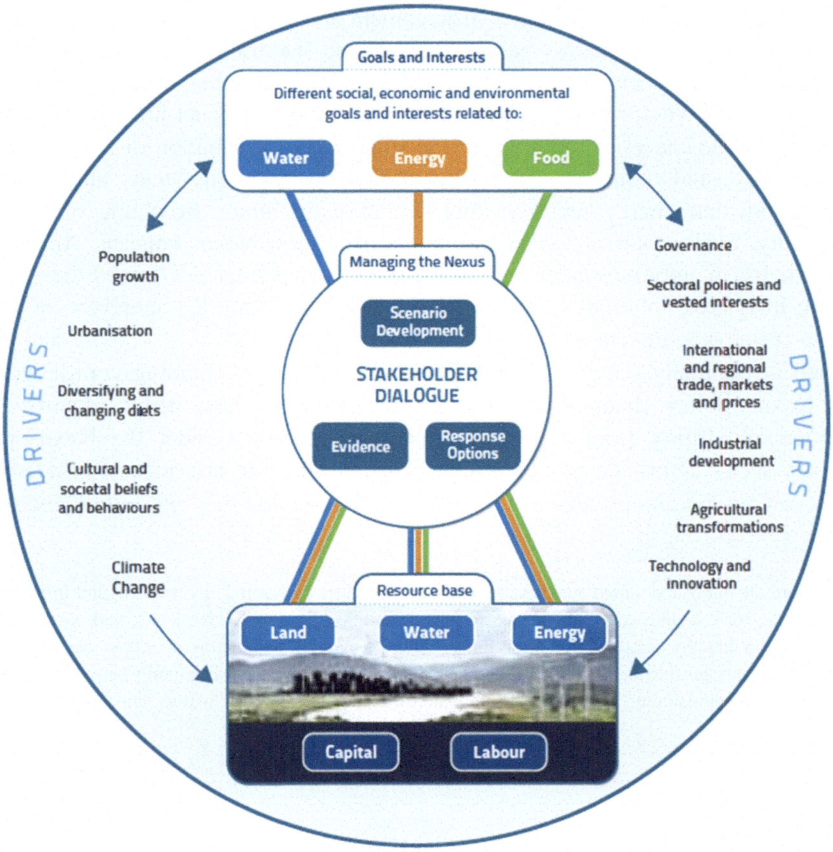

Figure 1.1 Presentation of the WEF Nexus (FAO).

Operationally, two components out of four could be addressed without harming, or at least minimizing, the impacts on the other two. This evidently becomes more complex with the addition of a third and fourth Nexus pillar, requiring the careful integration of numerous elements presented in Figure 1.1. Years of research and test implementation will need to be done before this approach is fully mainstreamed into cross-sectorial planning processes, but multi-sectorial policy analyses and dialogues addressing multi-sectorial issues and needs are already in practice and addressing concrete issues and challenges. For example, most climate change policies traverse and implicate multiple policy sectors such as water, agriculture, energy, and health: all of which have significant impact on ecosystems.

The key principles of the WEFE Nexus can be summarized as follows[2]:

- **Understand the interdependence of resources within a system across space and time** and focus on the whole system's efficiency rather than the productivity of individual components. This will provide integrated solutions that contribute to the sustainability of water, energy, food security policy objectives and to maintaining healthy ecosystems.
- **Recognize the interdependence between water, energy, food, and ecosystems** and promote rational and inclusive dialogue and decision-making processes and efficient use of these resources in an environmentally responsible way.
- **Identify integrated policy solutions to optimize trade-offs and maximize synergies across sectors** and encourage mutually beneficial responses that enhance the potential for cooperation between all components, and public–private partnership at multiple scales.
- **Ensure coordination across sectors and stakeholders** to enable synergies and increase solution sustainability.
- **Value the natural capital of land, water, energy sources, and ecosystems** and encourage governments and business to support the transition to sustainability, for example, using nature-based solutions.

1.3 THE CHALLENGES

The planet's natural resources are increasingly coming under pressure and suffering depletion with impacts on ecosystems in many places. Deficits in electricity are increasingly concentrated in sub-Saharan Africa; 789 million people – 85% in rural areas – lacked electricity in 2018. Since 2014, the global prevalence of undernourishment (chronic food insecurity) has remained virtually unchanged at slightly below 9%, but the total number of people going hungry has slowly increased for several consecutive years. Almost 690 million people were undernourished in 2019, up by nearly 60 million from 2014. Eliminating hunger alone will not ensure that everyone has access to sufficient nutritious food. An estimated 25.9% of the global population – 2 billion people – were affected by moderate or severe food insecurity in 2019, an increase of 22.4% from 2014. Approximately 2.2 billion people around the world still lack safely managed drinking water, including 785 million without basic drinking water. The population using safely managed sanitation services increased from 28% in 2000 to 45% in 2017. However, 4.2 billion people worldwide still lacked safely managed sanitation, including 2 billion who were without basic sanitation. An irreversible effect of human activity on the environment is species extinction and reduction in ground cover, which upsets the balance of nature and makes ecosystems more fragile and less resistant to disruptions.[3]

---

[2]Position paper on water, energy, food and ecosystems (WEFE) nexus and SDGs, JRC Technical Report, JRC114177, 2019.

[3]The Sustainable Development Goals Report 2020, https://unstats.un.org/sdgs/report/2020/

The core threats to the resources are: population growth; economic development; urbanization development; lack of transboundary cooperation; pollution; and climate change/variability. The environmental aspect is a reflection of these threats.

Population growth projections estimate that the global population will have grown up to around 9 billion by 2050 (7.3 billion in 2015). As a consequence, energy consumption is estimated to grow by 80% and food demand by 60% (IEA 2010, FAO 2012). Agriculture is already consuming 70% of all global freshwater abstractions.[4]

While **economic development** continues to move people from poverty into middle income living, this is resulting in increased demand for water, food, and energy as a consequence of consumption patterns. Even if many people are lifted out of poverty, the poorest sector of the population is growing even faster and the actual number of poor and vulnerable populations and inequalities are increasing. Increasing demand due to population growth and economic development combined with unsustainable production methods will put increased strain on the natural resources base that is unlikely to be reversed during the foreseeable future.

In addition, the **urbanization** phenomena can also constrain **WEFE** security and lead to ecosystem degradation as in the case of not well-managed urban development. The growth of urban population is estimated to add '*2.5 billion to the world's urban population by 2050, with almost 90% of this growth happening in Africa and Asia*'.

Some of these challenges have regional as well as national dimensions, and a more active coordination of management strategies and plans across sectors and borders is still needed to avoid unilateral exploitation and uneven competition for shared resources.

Climate change brings additional challenges to the WEFE Nexus because it impacts on resource availability (e.g., water resources access, quality, and quantity), related economic activities (agriculture and energy) and the overarching ecosystems component. The effects and impacts of climate variability need to be factored into resources planning and ecosystem services now more than ever before in short-, medium-, and long-term management strategies.

1.4 THE KEY ADVANTAGES

The WEFE Nexus aims to increase water, energy, food security and ecosystems without compromising ecosystems and destabilizing ecosystems services. In practical terms, the WEFE Nexus helps to improve understanding and analysis of the systems and interactions between the natural environment and human activities in these four Nexus pillars. This will develop more coordinated and sustainable management of natural resources, economic activities and strengthening ecosystems across sectors, levels, and scales.

Some key benefits arising from utilizing the Nexus are:

(1) *Exploitation of co-benefits to improve overall performance by:*

Increased resource productivity: Technological innovation, recycling of waste, waste reduction, and demand management can all contribute to improving utilization of available resources.

Waste becomes a resource: In particular, waste and manufacturing residues and by-products can become resources for various processes and a sustainable alternative to a traditional linear economy (take, make, use, dispose). This contributes to a circular economy concept which is restorative and generative (i.e., waste water to fertilizer, solid waste to fuel).

[4]The state of the world's land and water resources for food and agriculture (FAO 2011).

Demand management will be a natural outcome of WEFE Nexus analysis which will ensure security of the Nexus components. Demand can be analysed and become a catalyst oriented towards more efficient optimization of resources, waste management and recycling through and across uses and users, and limit the pressure on ecosystems. Two examples are the use of wastewater in agriculture and the use of renewable energy for pumping water for irrigation and domestic uses.

Alternative technology and practices: This is based upon departure from technologies and practices which produce excessive waste and/or have negative impacts on health and the environment. For example, wood fuel can be optimized or replaced by other cooking methods (e.g., solar) to reduce the related demand for wood, the wasteful practice of charcoal production and limit impacts on the ecosystem and family/community health. Introducing alternative sanitation measures such as dry or composting latrines helps protect water resources and recycles waste for agricultural purposes, while reducing the water demand from water-based sanitation services/systems. Wastewater and greywater sludge can also be recycled or re-purposed for potable water, energy, and fertilizer. The Nexus approach can bring these three independent alternatives together into a complementary and more efficient technologies and practices while minimizing impacts on ecosystems and the environment.

(2) *Streamlining development and improving resilience*

Benefits from productive ecosystems: The Nexus approach also aims to preserve ecosystems which not only have intrinsic value (e.g., maintaining biodiversity), but also provide multiple services; increase overall the ecosystems benefits. In taking advantage of natural infrastructure and soft path solutions, Nexus-based solutions which integrate ecosystems can complement man-made 'hard' infrastructure and end-of-pipe solutions and eventually be able to deliver certain services more efficiently (e.g., improved water quality, urban green spaces for flood management, small-scale food production and micro-climate control).

Poverty alleviation: The Nexus integrative approach, because it is not driven by a fixed sector and its defined stakeholders, can have greater efficiency and impact in advancing the basic services of water supply, alternative energies, and food, and also helps to strengthen and protect ecosystems and maintain healthy living environments. This sharing of capacities helps greatly to extend the reach of benefits, ultimately to the poorest populations who often derive their livelihood directly from the local ecosystem and more vulnerable to ecosystem damage.

Climate change: Considered that, climate change, climate variability, mitigation and adaptation are included in the dialogues, the Nexus approach creates a wider transparency to planning processes, making them more accessible and easier for stakeholders to understand and accept. The Nexus approach, with its wider capture of stakeholders and key sector actors, provides a greater potential contribution to enhancing resilience and to reducing disaster risk (e.g., water floods can be stored in the dam and released for agricultural and energy production purposes later).

(3) *Stimulating policy coherence and multipurpose investments*

Governance, institutions, and policy coherence: The Nexus policy dialogues (consultation) embedded in the Nexus approach require a high degree of collaboration between, populations/communities, policy sectors and associated institutions (including the local administrations and traditional ones). As such, benefits can be achieved on social, economic, and environmental values because sustainable solutions and policy coherence improvements come from well-structured consultations and collaborations across the actors/stakeholders to build shared objectives and resources. The existing institutional management frameworks such as national, transboundary or regional agencies are governance infrastructures that must ultimately support and integrate the Nexus processes and its implementation.

SDGs support: A single sector approach will not contribute to an integrated and consistent SDGs achievement, limiting the trade-off. Analysis of the content of the four pillars of the WEFE Nexus in the SDG charter indicates that the WEFE Nexus components are present in 14 of the 17 SDGs and are therefore highly relevant in terms of working towards the SDGs' implementation (see Section **XX** WEFE Nexus and SDGs)

Stimulate development through multipurpose investment: The Nexus approach can help stimulate investment in sustainable infrastructure and help avoid preventing development by only investing in a single sector. Pricing of ecological services can also help direct investment in more sustainable development projects by combining built and natural infrastructure.

1.5 THE PROCESS

Managing the WEFE Nexus is a consultative process involving key stakeholders contributing to and agreeing on responses to the challenges being faced. Consultation is particularly crucial when implementing the WEFE Nexus because of the need to collaborate across traditional thematic silos. It is the main vehicle for reaching consensus as it relies on the following principles:

- Bring together stakeholders from different sectors, different spheres of government and countries, and different levels;
- Link directly to ongoing and emerging processes;
- Develop a shared understanding of issues, objectives, and scenarios; and,
- Help achieving concrete agreements on multi-sectorial and multi-scale strategies to design concrete intervention projects/solutions in view of achieving the SDGs.

Figure 1.2 presents the steps in the Nexus process from evidence to assessment and policy dialogues; leading ultimately to implementation. They are explained in the following sections.

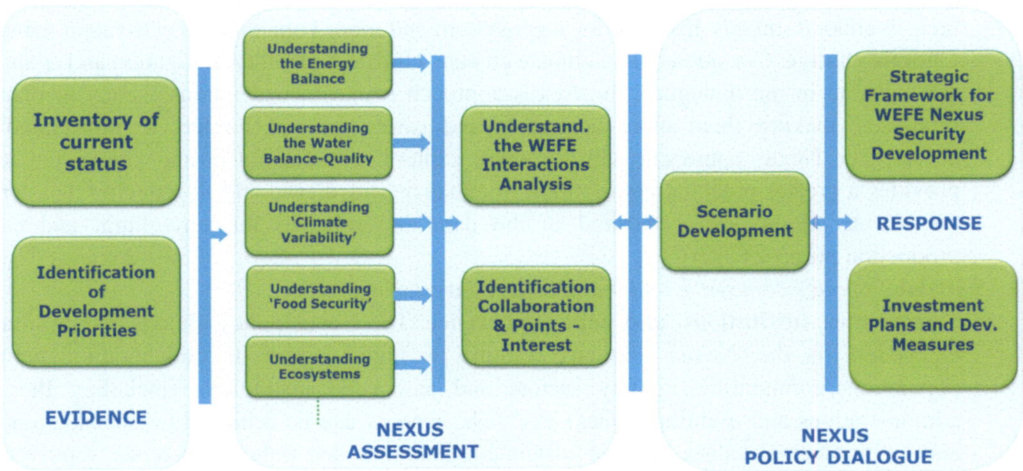

Figure 1.2 Nexus dialogue process (JRC Technical Report, Position Paper 2019).

Evidence: Nexus assessment initially requires data and knowledge collection to identify the linkages between the water, energy, and food systems, and the impact that changes to these systems can have on the ecosystems and livelihoods. Data and knowledge collection will have to revolve around natural resources availability, socio-economic context and trends, and policies implemented in relevant sectors. To simplify the complex field of applicability, information and data have to be adapted to the context, linked to the policy priorities and needs, and to the complexities of the environment. The information can be both qualitative and quantitative depending on the case in point. The implementation of this task will consider: (a) that there is often a lack of good, accurate, harmonized, and up-to-date knowledge, data, and information; (b) asymmetrical access to information; (c) that available data and knowledge in transboundary basins may differ greatly in terms of level of aggregation, scale, accuracy, reliability, etc. not only between countries sharing resources but also institutions. Sharing of information, with as open access as possible, between institutions involved in data and knowledge collection and management needs to be encouraged as information on critical conditions in one sector impact the other Nexus components. This is particularly important in the case of transboundary waters, where various actors are involved in the management of the resources and the respective information and data.

Nexus assessment: The Nexus assessment process will lead to improved understanding of the interactions between the different sectors and their potential impacts on the environment. Nexus assessment outputs shall be a concrete set of optimal solutions (recommendations) coming from the integrated and collaborative models and evaluations. These scientifically and technically based outputs will then feed the policy dialogues concerning the predefined priorities and needs. What can be achieved in a Nexus assessment depends on various factors: the context, the issues, the actors, the capacities involved, the constructiveness of the dialogue, the availability of information (data and knowledge …), and the political will.

1.5.1 Policy dialogue

Scenario development: This shall identify short-/medium-/long-term effects of possible Nexus intervention or the application of new policies on the natural environment and society at various time and spatial scales. It includes the estimation of potential benefits on the Nexus components and must highlight impacts on the SDGs, and lead to a shared vision of water, energy, and food security in a sustainable environment (healthy ecosystems). The scenario can be developed by applying the existing/validated/calibrated tools and models in the various sectors (for instance, life-cycle assessment, energy or agro-hydrological modelling). These are applied to the different Nexus components by combining them. The outputs of one component/model can become inputs in another one within the framework of an iterative assessment process.

Response options: Consensus will be reached on specific policy options and trade-offs that address planned interventions in an open, participatory, and inclusive dialogue during stakeholder consultation. The response options need not be built from comprehensive quantitative and qualitative data collection but can be drawn out of qualitative analysis immediately after the evidence gathering phase.

These four elements can be iterative with the objective of integrating new key issues or evidences arising out of the various phases, along the analytical process.

2

THE WEFE Nexus in the context of EU development policy

Recent key EU development policy framework documents include 'An Agenda for Change (2011)', the 'New European Consensus on Development (2017), the EU External Investment Plan (2016), 'A Global Strategy for the European Union's Foreign And Security Policy' (2016), and various sector development policies. For example, the INTPA water Nexus policy[1] states that: '*...challenges......can be addressed through ensuring a better management of linked resources. One example is to reconcile the competing needs of water for energy and water for agriculture while securing needs for underpinning ecosystems. This needs to be done in an integrated, transboundary and equitable manner, and enhancing cooperation across borders. Especially in the context of the holistic approach adopted by the post-2015 agenda, the EU will need to have goals and targets in specific areas while at the same time ensuring that it does not create a 'set of silos' of completely separate goals*'. A presentation on the EU water development policy (2012) explicitly mentions the water–energy–food security Nexus under its framework for action. The INTPA Sustainable Energy Handbook (2016) has a specific Module 2.4 on the water–energy–food Nexus.

The above-mentioned New European Consensus on Development 'Our World, Our Dignity, Our Future' is a Joint Statement by the Council and the Representatives of the Governments of the Member States meeting within the Council, the European Parliament, and the European Commission. It serves as the overarching EU development policy framework and from now onwards is referred to as the 'Consensus' for short. The Council of Europe Press Release Joint Statement on the adoption of the Consensus states that: '*Our new approach to development is based on the 'five Ps' of the 2030 Agenda: People, Planet, Prosperity, Peace, and Partnerships.* ***Recognising the interlinkages between the SDGs, we will pay attention to actions that meet multiple goals in a coherent way****. We will* ***work across policies and sectors to boost synergies addressing a range of cross-cutting elements*** *to accelerate transformation. Based on the* ***principle of policy coherence for development****, development objectives will be fully taken*

[1]https://ec.europa.eu/europeaid/sectors/infrastructure/water-and-sanitation/water-Nexus_en

into account across EU policies that are likely to affect developing countries. Policy coherence will be ensured to contribute to the achievement of the SDGs by partner countries'.

Paragraph 19 of the Consensus document states: '*The implementation of the 2030 Agenda requires comprehensive national sustainable development strategies that* ***factor in the SDGs and their interlinkages****. When planning and implementing development cooperation, the EU and its member states will pay* ***particular attention to such interlinkages and to integrated actions*** *that can create co-benefits and meet multiple objectives in a coherent way*'. This and many other key policy statements in the Consensus directly or indirectly provide the EU policy context for a WEFE Nexus approach in EU development cooperation.

It is also noted that the July 2018 UN High Level Policy Forum (HLPF[2]) focussed on the themes in SDGs 6, 7, 11, 12, 15, and 17 which was highly relevant from a WEFE Nexus perspective. Related to its participation and contributions to the HLPF, the European Commission stated[3] that 'Investment in these SDGs will yield significant co-benefits for the whole 2030 Agenda, which will only be achieved through an integrated, holistic approach'. The Ministerial Declaration from the HLPF stated that: 'While our 2018 review emphasizes Sustainable Development Goals 6, 7, 11, 12, 15, and 17, the integrated, indivisible, and universal nature of the goals makes it essential that we pay particular attention to leveraging synergies and co-benefits across all dimensions of sustainable development'. The HLPF President's summary[4] contains a reference to the 'land–food–water–energy–climate Nexus' and mentions that many HLPF speakers expressed their increased understanding of how progress could be leveraged through addressing the many interlinkages between the SDGs. In connection with the HLPF, the European Commission stated[5] that it, together with its member states, will prepare a Joint Synthesis Report on the implementation of the European Consensus on Development for the 2019 session of the HLPF, and that it has established a high-level multi-stakeholder platform[6] on the implementation of the SDGs.

2.1 SUBSTANTIVE POLICY FOCUS IN THE CONSENSUS RELEVANT TO THE WEFE NEXUS AND RELATED LINKS TO THE SDGs

The table below identifies key areas of policy emphasis in the Consensus that are relevant to the WEFE Nexus – structured according to the 5 Ps. For each of the highlighted Consensus policy quotes, reference is made to the relevant SDGs, illustrated by the relevant SDG icons. To facilitate an overview, many of the quotes from the Consensus are not given as full sentences – but paragraph numbers (para no.) are used to indicate the location of the full text in the Consensus document.

[2]The theme of the overall HLPF was 'Transformation towards sustainable and resilient societies' and the specific focus was SDG 6 (water), SDG 7 (sustainable energy), SDG 11 (urban development), SDG 12 (sustainable consumption and production patterns), SDG 15 (ecosystems), and SDG 17 (partnerships).

[3]The EC INTPA statement re HLPF: https://ec.europa.eu/europeaid/news-and-events/un-high-level-political-forum-eu-and-un-review-progress-towards-achieving-2030_en also noted that the voluntary national reviews (VNRs – a mechanism designed to assess the progress of each UN Member State in achieving the SDGs) would be considered at the HLPF. The 153 available VNRs can be found in the VNR database. The updated 2018 VNR guidelines emphasize examining policy coherence and interlinkages between SDGs, but many VNRs have limited focus on these aspects.

[4]https://sustainabledevelopment.un.org/content/documents/205432018_HLPF_Presidents_summary_FINAL.pdf

[5]http://sdg.iisd.org/news/stakeholders-launch-sdg-report-outlining-vnr-best-practices-and-recommendations/

[6]https://ec.europa.eu/info/strategy/international-strategies/global-topics/sustainable-development-goals/multi-stakeholder-platform-sdgs_en

EU policy Emphasis Relevant to a WEFE Nexus Approach	Para no.	Consensus Policy Quote	Related SDGs
People:			
The policy of supporting interventions addressing **food security**.	24	*'….make coordinated, accelerated and **cross-sectoral efforts to end hunger**, increase the capacity for diversified local and regional food production, ensure food security and nutrition and enhance the resilience of the most vulnerable…'; and 'addressing all forms of malnutrition….through support for basic services in health, nutrition, water, sanitation and hygiene, and social protection'.*	
EU policy of supporting **access to land, food, water and energy**.	25	***'The EU and its Member States will support the poorest communities in improving access for all to land, food, water, and clean, affordable, and sustainable energy while avoiding any damaging effects on the environment.*** *They will promote policy initiatives and support partner countries in* ***planning and implementing an integrated approach to concretely address the most relevant interlinkages between land, food, water, and energy.'***	
EU policy on **integrated water management**.	26	*'… support sustainable and **integrated water management**, as well as more efficient use of water and water recycling, including a more strategic approach to regional development and integration'.*	

(*Continued*)

EU policy Emphasis Relevant to a WEFE Nexus Approach	Para no.	Consensus Policy Quote	Related SDGs
EU policy on **migration**.	40	*'Addressing **migration cuts across many policy areas, including development, good governance, security, human rights, employment, health, education, agriculture, food security, social protection and environment, including climate change.** Through the Partnership Framework approach the EU and its Member States will address in a comprehensive manner the multiple aspects of migration and forced displacement…'.*	
Planet:			
EU policy emphasis on **environmental sustainability.**	43	*'The EU and its Member States will promote resource efficiency and sustainable consumption and production, including the sustainable management of chemicals and waste, with a view to decoupling economic growth from environmental degradation and **enabling the transition to a circular economy**', and …. 'help to build capacity to **mainstream environmental sustainability, climate change objectives and the pursuit of the green growth** into national and local development strategies'.* And para#44: *'The EU and its Member States will promote the use of **natural capital accounting**'.*	

EU policy on **climate change** indicates cross-sectorial and coordinated action.	45	*'…. will implement the 2030 Agenda and the Paris Climate Change Agreement through coordinated and coherent action and will maximise synergies. They will support national strategies, including cross-government planning and programming, which* ***promote resilience, reduce climate risk and contribute to emission reduction, consistent with the implementation of Nationally Determined Contributions (NDCs)****, taking into account the challenges faced by developing countries, particularly Least Developed Countries (LDCs) and Small Island Developing States (SIDS)'…… including by supporting multi-stakeholder platforms. The legally binding character of the Paris Agreement and the requirement to prepare NDCs can also give impetus to national development planning in the context of the 2030 Agenda'.*	
EU policy on **sustainable energy**.	the box after para. 5	*'Investment in* ***sustainable energy can ensure and increase access to clean water, clean cooking, education, and healthcare, and can also create jobs and support local businesses in an environmentally friendly manner****. The EU and its Member States will pursue three interlinked key objectives: addressing the lack of energy access; increasing energy efficiency and renewable energy generation to achieve a sustainable balance between energy production and consumption; and contributing to the global fight against climate change in line with the Paris Agreement….'*	

(*Continued*)

EU policy Emphasis Relevant to a WEFE Nexus Approach	Para no.	Consensus Policy Quote	Related SDGs
Prosperity:			
EU policy on inclusive and sustainable growth and its emphasis on **value chains**.	52	*'The EU and its Member States will promote inclusive and sustainable economic growth and help developing countries* ***adopt growth models that take account of resource scarcity and climate change action****. This includes promoting* ***sustainable value chains****….'*	
EU policy emphasis on the **private sector's** role.	53	*'…****promoting private sector accountability****, in areas with significant transformation potential for sustainable development. This includes* ***sustainable agriculture, safe and clean energy, integrated water resource management, resilient infrastructure, health, sustainable tourism, green and circular economy****….'.*	

EU policy on **sustainable agriculture** clearly indicates the importance of the links with water and ecosystems.	55	*'The EU and its Member States will support improvements in* ***governance relating to sustainable forest management, participatory rangeland management, and to equitable access to land tenure****, particularly for women, respecting the rights of local populations and of indigenous peoples, including customary land use and access to water'. And Para #56: '…support agro-ecological practices and actions to reduce post-harvest losses and food waste, as well as to protect soils, conserve water resources, halt, prevent and reverse deforestation, and maintain biodiversity and healthy ecosystems'.*	
EU policy on **sustainable urbanization.**	60	*'The EU and its Member States will seek to* ***boost the potential of cities as hubs for sustainable and inclusive growth and innovation****, taking account of their wider rural communities and of balanced regional development'; '….promote inclusive, balanced, integrated territorial and urban policies, and multi-level governmental coordination, forging stronger links between rural and urban areas….. build cities' resilience to shocks and harness opportunities for a low-emission and climate-resilient economy'.*	

(*Continued*)

EU policy Emphasis Relevant to a WEFE Nexus Approach	Para no.	Consensus Policy Quote	Related SDGs
Peace:			
EU policy related to conflict and fragility underlines the importance of an **integrated approach to poverty eradication**.	64&65	*'**Poverty, conflict, fragility, and forced displacement are deeply interlinked and must be addressed in a coherent and comprehensive way including as part of the humanitarian-development Nexus**. The EU and its Member States will address their root causes at all levels, ranging from exclusion, inequality, food insecurity, human rights violations and abuses, impunity, and the absence of the rule of law to environmental degradation and climate change'. '**The prime focus of development cooperation remains poverty eradication** in all its dimensions, and there will be no diversion of efforts from that goal'.*	
EU policy on climate change **resilience and adaptation** and the related needs for gap analyses.	70	*'The EU and its Member States will increase their efforts to **build resilience and adaptability to change** consistent with, inter alia, the Sendai Framework for Disaster Risk Reduction 2015–2030 and the Paris Agreement on Climate Change…. Will build risk assessments and gap analysis into their development cooperation programmes'.*	

Partnership:			
EU development policy underlines **working more closely with all other relevant actors** in the implementation of the 2030 Agenda.	83	*'Stronger partnerships are at the heart of the EU's approach to SDG implementation. The EU and its Member States will* ***work more closely with all other relevant actors to promote the implementation of the 2030 Agenda****'.* And para. 87: '*The successful implementation of the 2030 Agenda also requires forging stronger partnerships beyond governments. The EU and its Member States will* ***expand partnerships with the private sector, civil society, including trade unions and employers organisations, multilateral and regional organisations, academia, diasporas. and other relevant stakeholders****.'*	

With these policy emphases in mind, we can summarize the concepts listed above before moving onto the consensus policy pertaining to the EU's policies with WEFE Nexus. The European Union (EU) has a multi-disciplinary approach when it comes to climate and sustainability. Because of this, the policies relevant to WEFE come from a variety of aspects, ranging from the policies on climate change, to sustainable agriculture or the integrated approaches to poverty eradication. The fact that the WEFE Nexus considers both the environment and the human–ecosystem interactions, to evaluate the viability for healthy communities and ecosystems, makes it very aligned to the EU's comprehensive sustainability approach and policy project. Therefrom, the strong EU's interest in properly developing the WEFE Nexus approach. Based on the above considerations, an evaluation of the WEFE Nexus application to the context of consensus policy is provided in the following sections.

2.2 THE WEFE NEXUS APPROACH IN THE CONTEXT OF CONSENSUS POLICY

The New EU Consensus on Development document sets out strategic approaches to improve EU impact in the implementation of the 2030 Agenda. Some of these Consensus policy statements point to concrete actions that can be entry points for facilitating the implementation of a WEFE Nexus approach at a strategic level. This section identifies the key Consensus policy statements, highlighting their relevance and potential synergies with specific WEFE Nexus entry points (for simplification purposes, no reference to SDGs is added at this stage). This overview analysis does not address the EU methods in detail nor the instruments for concrete operationalization, which needs to further development.

Entry Point for Enabling a WEFE Nexus Approach	Para No.	Consensus Policy Quote
Using the SDGs to catalyse policy coherence in implementing the EU's Global Strategy[7] that provides an overall vision for the EU's engagement with the world.	9	*'The EU Global Strategy sets out a vision for the EU's engagement in the world, through a range of policies. It highlights the important role of the 2030 Agenda, which has the potential to trigger the necessary transformation in support of EU values and the objectives of EU external action. The SDGs will be a cross-cutting dimension of all the work to implement the EU Global Strategy.'*

(*Continued*)

[7]This Strategy (explained in Section 4) states that *'The Sustainable Development Goals also represent an opportunity to catalyse … coherence. Implementing them will generate coherence between the internal and external dimensions of our policies and across financial instruments. It allows us to develop new ways of blending grants, loans, and public–private partnerships. The SDGs also encourage the* ***expansion and application of the principle of policy coherence*** *for development to other policy areas, and encourage joint analysis and engagement across Commission services, institutions, and Member States'* and *'We will also* ***support governments in devising sustainable responses to food production and the use of water and energy*** *through development, diplomacy, and scientific cooperation'.*

Entry Point for Enabling a WEFE Nexus Approach	Para No.	Consensus Policy Quote
Apply the EU's development effectiveness principles[8]: the EU and its Member States could address the WEFE Nexus in relevant **policy and development cooperation forums** to ensure that this approach is implemented in a coherent manner where relevant: in joint programming, public–private engagement, in the efforts to 'leave no-one behind' and in the transparency of development cooperation.	73	*'This includes improving effectiveness and impact* ***through greater coordination and coherence,*** *by applying the development effectiveness principles, and by delivering development cooperation as one part of the overall internal and external action to promote the implementation of the 2030 Agenda'.*
More specifically, use opportunities to address the Nexus in joint analyses and joint interventions in order to enhance **Joint Programming** at country level.	75	*'…. The EU and its Member States will work together to develop strategic responses grounded in shared knowledge, added value, lessons learned and* ***joint analysis of the country context, including poverty and sustainability****…' and in parag. 76: 'This approach will help pool resources, reduce fragmentation and boost effectiveness' and in parag. 77: Joint implementation is a way of promoting more coherent, effective, and coordinated EU support based on shared objectives in selected sectors or on specific* ***cross-sectorial themes*** *tailored to the country contexts'.*
Make use of opportunities for strategic policy dialogue about the benefits of a Nexus approach in preparing and implementing EU **budget support**.	81	*'Budget support, when applicable and with those willing to participate, will be used to strengthen partnership,* ***political dialogue****, country ownership and mutual accountability with developing countries…'*
Take opportunities to advance a Nexus approach through the EU's major leverage in **blending** as a means to mobilize additional finance to implement the 2030 Agenda and when implementing the European External Investment Plan (EIP).	82	*'Blending grants and loans, as a way to leverage additional private finance, is another important means to implement the 2030 Agenda.* ***Blending covers all regions of EU external cooperation in sectors including energy, transport and water*** *infrastructure, support for small and medium enterprises, social sectors and the environment. Blending is a major component of the European External Investment Plan'.*

(*Continued*)

[8]https://ec.europa.eu/europeaid/policies/eu-approach-aid-effectiveness_en

Entry Point for Enabling a WEFE Nexus Approach	**Para No.**	**Consensus Policy Quote**
Advance the Nexus thinking in **capacity development** efforts as EU engages more closely with all relevant national partners, recognizing the important role of support for comprehensive and inclusive **planning rooted in national and sub-national development strategy** programmes and budgets and the related national monitoring frameworks. Support for strengthening of the aforementioned national voluntary reviews (NVRs) of SDG implementation could be a priority.	85	*'The EU and its Member States will support* ***capacity building*** *for nationally owned monitoring frameworks, quality data collection, disaggregation and analysis, including through digital monitoring tools and for* ***policy coherence for sustainable development****'.*
Use opportunities for more concerted multilateral action on advancing a Nexus approach because the EU is strengthening its **partnerships with other multilateral organizations**. The above-mentioned UN HLPF annual events may offer good opportunities for demonstrating the advantages of a Nexus approach in the implementation of the SDGs, and the HLPFs are good opportunities to assess the NVRs in a multilateral setting with national high-level delegations and multilateral development partners. The European Development Days (EDDs) and similar events attended by the wider international development community may also provide opportunities to advance concerted Nexus action.	90	*'The EU and its Member States will* ***strengthen their partnerships with multilateral organisations****, including the United Nations system, the International Monetary Fund, the World Bank Group, regional development banks, the G7, the G20, the OECD and other regional and multilateral institutions. They will encourage them to align their strategic planning and operational activities with the 2030 Agenda and foster mutual and coordinated support in implementation thereof, in full alignment with national sustainable development strategies'.*

(*Continued*)

Entry Point for Enabling a WEFE Nexus Approach	Para No.	Consensus Policy Quote
Take opportunities to demonstrate innovation and examples of the advantages that a Nexus approach can offer as the EU places emphasis on fostering **innovative engagement with more advanced developing countries**. Facilitating the exchange of the lessons, best practices from emerging economies, and using the 'power of the example' could have strong influence on weaker developing economies and fragile countries, and there may be opportunities for 'leapfrogging' to apply solutions that benefit from Nexus approaches. This could also involve new partnerships with centres of excellence that have demonstrated successful Nexus solutions.	Box after para. 95	*'The EU and its Member States will develop new* ***partnerships with more advanced developing countries*** *in order to promote the implementation of the 2030 Agenda through a broader range of cooperation…. These new partnerships will* ***promote the exchange of best practices****, technical assistance, and knowledge sharing… will work with these countries to promote* ***South-South and triangular cooperation****'.*
Work with partner countries to enhance the enabling environment for a Nexus approach as the EU will work with partner countries to promote **sound policy environments for implementing the 2030 Agenda**. It is widely recognized and well-documented that the SDGs are interconnected and that working closely with all relevant stakeholders is beneficial. This includes the private sector, which is key to achieving the SDGs. This will also include a 'whole of government approach' working across traditional development sectors to strengthen the enabling environment for more concerted action.	99	*'… will promote policies linking public and private pro-development action and an* ***enabling environment for inclusive sustainable growth*** *and its equitable distribution through national budgets. They will plan their development cooperation around strengthening countries' own capacities to implement the 2030 Agenda'.*

(*Continued*)

Entry Point for Enabling a WEFE Nexus Approach	Para No.	Consensus Policy Quote
Support a Nexus/cross-sectorial approach in pursuing **policy coherence for development** (PCD), which is key to achieving the SDGs. The PCD is a key element of EU Consensus policy and there are many opportunities for Nexus thinking in the dynamic policy environment for development cooperation and in strategic policy dialogue with partner countries. This is the case, for example, in budget support and programmes related to thematic areas such as climate change that cut across traditional sector policy areas.	109	*'The EU and its Member States reaffirm their* ***commitment to Policy Coherence for Development...*** *The 2030 Agenda provides new impetus for the EU and its Member States to formulate and implement mutually reinforcing policies'.* And para#110: *'The Consensus will guide efforts in applying PCD across all policies and all areas covered by the 2030 Agenda, seeking synergies, notably on trade, finance, environment and climate change, food security, migration and security'.* And para#112: *'The EU and its Member States will moreover strengthen their dialogue with partner countries on policy coherence and support partner countries in their own efforts to put in place enabling frameworks for policy coherence for sustainable development'.*
Pursue Nexus thinking **in value chains and other approaches** that link public and private actors at different levels of governance in development programmes, recognizing that cross-sectorial and more holistic 'whole of government' approaches are needed to achieve interconnected SDGs.	111	*'Sustainable development* ***requires a holistic and cross-sector policy approach*** *and is ultimately an issue of governance....the EU and its Member States will therefore* ***promote whole-of-government approaches*** *and ensure political oversight* ***and coordination efforts at all levels for SDG implementation.*** *Ongoing EU* ***action towards sustainable global supply chains****, such as in the timber and garment sectors,* ***illustrate the added value of pursuing a coherent approach****'.*
Facilitate a Nexus approach by aligning the theory of change of EU development cooperation and its results frameworks with SDG indicators. It is EU policy that the **EU's reporting systems** will be made consistent with the 2030 Agenda indicators, and as described in the foregoing, coordinated, cross-sectorial approaches are inherent in the SDGs. Consequently, this can be an effective means of supporting a Nexus approach.	119	*'The EU and its Member States* ***will integrate the 2030 Agenda and support the use of SDG indicators*** *to measure development results at country level.'*

(*Continued*)

Entry Point for Enabling a WEFE Nexus Approach	**Para No.**	**Consensus Policy Quote**
Make use of the planned **joint synthesis reports** on the Consensus to assess how the Consensus policy has facilitated a WEFE Nexus approach and where further actions is needed. As mentioned above the EU is planning to submit a joint synthesis report to the 2019 HLPF and this offers a very good opportunity to demonstrate where the Consensus has facilitated Nexus approaches and where further action is needed.	120	*'The EU and its Member States will produce a **joint synthesis report** on the Consensus on Development, including the impact of their actions in support of the 2030 Agenda in developing countries, as a contribution to EU reporting to the UN High Level Political Forum (HLPF) when meeting at Head-of-State level every four years'.*
Similarly, the EU should make use of the **2024 mid-term assessment** of the Consensus to address progress on policy coherence and Nexus approaches. A plan of action could be developed to elicit examples and lessons learned over the coming years on how Consensus areas of policy emphasis and its (limited) emphasis on modalities has facilitated success stories on development outcomes and impact facilitated by a WEFE Nexus approach.	123	*'A **mid-term assessment** of the implementation of this Consensus will be carried out by 2024.'*

3

Conclusions drawn from EU-WEFE Nexus consensus policy

Analysing the policy components, there are some general conclusions to take away from the WEFE Nexus approach in the context of consensus policy. Primarily, it is worth noting the alignment in the necessity for WEFE Nexus and the EU to have cooperation forums and joint programming. EU procedures, like the WEFE Nexus approach, needs to rely on multi-disciplinary discussion among different organizations and institutions, agreeing on policy coherent and a joint plan of action. EU support to the WEFE Nexus programme leverages EU institutional capacities in blending and for the 2030 Agenda's implementation. The Nexus allows for a more inclusive strategy and action planning both at national and subnational level programme implementation, while allowing stronger cooperation with multinational and international organizations (such as the EU and UN agencies). For multi-sectoral programmes, the Nexus approach can provide a structure for smoother reporting. Finally, it enhances the value chains and links between the public and private actors and levels of governance in the project's location.

3.1 WEFE NEXUS AND THE GREEN DEAL

With the new Commission in place, the political frameworks have evolved even further. The current pandemic gives us a glimpse of what could be in store in the future in the absence of ambitious action. Studies show that climate change will increase the number and frequencies of crises like the current pandemic, but also of conflicts, migration, and natural disasters.

Europe–Africa relations are facing a double challenge – the coronavirus disease-2019 (COVID-19) pandemic puts social and economic systems under strain at a point when the consequences of the climate crisis are being increasingly felt on both continents. Within Africa and Europe, debates have started about recovery measures to address the pandemic's short- and medium-term socio-economic consequences. A key question in these debates is how to 'build back better' and use the crisis to promote green transitions and move towards more sustainable development pathways.[1]

Green transitions have the potential to support Africa–Europe cooperation by combining the climate agenda with an innovative socio-economic project for jobs creation and sustainable growth. Green transitions can be a fruitful area for cooperation because common interests and interdependencies between both continents are high.

According to an analysis of the European Parliament, the proposals for a 'Comprehensive Strategy with Africa' place more emphasis on the economic aspects of sustainable development compared to social and environmental concerns. More precisely, in the strategy, the EU prioritizes the formal, productive and technology sectors as well as climate mitigation at the expense of the informal sector, human development, agriculture and climate adaptation.[2] There is a strong emphasis from the EU on green energy (transitioning away from fossil fuels) and on agriculture; both for food security in challenging climate extremes and for agricultural economic activity. In response to the COVID-19 pandemic, African leaders have called for a green stimulus programme, focusing investments on food production, water management and infrastructure with a view to addressing the socio-economic crisis resulting from COVID-19 and climate crisis at the same time.[3]

That said, on 19 November 2018, the Foreign Affairs Council adopted new Conclusions on EU Water Diplomacy and made the case for making the link between water, security and peace, including the potential of water as an instrument for peace. This was followed in January 2021 by The Council adopting conclusions on 'Climate and Energy Diplomacy – Delivering on the **external dimension** of the **European Green Deal**'. In its conclusions the Council recognizes that climate change is an **existential threat to humanity**. It notes that **global climate action still falls short of what is required** to achieve the long-term goals of the Paris Agreement and the 2030 Agenda for Sustainable Development. Therefore, there is a growing recognition around the Green Deal implementation that Green Economy in itself is not sufficient to achieve a transition to development, but that multiple sectors are strongly implemented as well as the social and human dimensions of policies, human capital and capacity, research and innovation and bridging the digital divide for better data and information for multi-sectoral resources and services management as well as raising up the capacities of individuals to engage in green economies.

The WEFE Nexus dialogues and pilot implementations has already begun to address these issues and has been accumulating knowledge, expertise and an increasing number of partners and stakeholders towards this end. In terms of addressing climate change challenges, the key WEFE Nexus sector activities are identifying sector challenges and strategies addressing human, social and economic capital gaps and needs and are well-placed to not only build on this experience but provide a framework for the EU support to the AU Green Transition. In addition, the active engagement of AU Partners in WEFE Nexus activities ensures that the pursuit of an EU-supported Green Transition remains essentially one to be developed within an African Agenda.

---

[1]Green Transitions in Africa–Europe relations: What role for the European Green Deal? (ETTG, April 2021).

[2]European Parliament's Committee on Foreign Affairs (2020): A comprehensive EU strategy for Africa. Political dialogue: Governance, security and migration (https://www.europarl.europa.eu/RegData/etudes/BRIE/2020/603507/EXPO_BRI(2020)603507_EN.pdf)

[3]African leaders endorse plan from Global Center on Adaptation for Africa to build climate resilience into recovery from COVID-19 pandemic (2020) (https:// www.prnewswire.co.uk/news-releases/african-leaders-endorse-plan-from-global-center-on-adaptation-for-africa-to-build-climate-resilience-into-recovery-fromCOVID-19-pandemic-835257673.html); Global Center on Adaptation (2020): Integrated responses to building climate and pandemic resilience in Africa. Policy Brief, 22 May 2020 (https://gca.org/reports/integrated-responses-to-building-climate-and-pandemic-resilence-in-africa/)

4

Case studies

The current chapter presents a selection of contributions addressing both cross-cutting issues and specific case studies relevant to the operationalization of the WEFE Nexus concept, the objective being to stress major challenges and benefits deriving by its adoption.

The majority of the contributions come from the proceedings of the 2018 WEFE Nexus Conference. With their global geographic scope, they inspired the discussions held by experts and, together with the Position paper, largely influenced the conclusions of this publication.

Additional case studies from more recent WEFE Nexus projects in Africa include:

- The ACEWATER2 project (2016–2020) and NEXUS Dialogues – Scientific Component project both funded by the European Union through DG INTPA and managed by the JRC in collaboration with UNESCO. The project addressed both scientific research on WEFE Nexus assessment and Human Capacity Development, in most of the sub-Saharan Africa, with the active involvement of a large base of continental, regional, and local stakeholders, including Institutions as AU, AMCOW (African Ministers' Council on Water), RECs (Regional Economic Communities), RBOs (River Basin Organizations), and more than 20 CoEs (Centers of Excellence) from Western, Central Eastern and Southern Africa.
- The Senegal WEFE nexus project (still ongoig), funded by DG-INTPA and implemented by the JRC in collaboration with the AICS (Italian Cooperation and Development Agency), pursuing the scientific objectives of: (i) reinforcing the technical and scientific knowledge on key phenomena occurring in the Senegal river basin, in collaboration with local and regional stakeholders; (ii) proposing sustainable management policies, consistently with the established institutional and legislative frameworks; (iii) supporting the OMVS (Organisation pour la Mise en Valeur du fleuve Sénégal) in measures operationalization over select pilot areas.

The first section of papers addresses the cross-cutting key challenges and major bottlenecks due to data availability and accessibility, implementation methodologies and tools, and blended financing. The second section of papers is organized by geographic area of interest, namely sub-Saharan Africa (Western, Central-Eastern, and Southern Africa), MENA region (Middle East and North Africa) and Central Asia.

WEFE Nexus cross-cutting issues

Water–energy–food Nexus: methodologies and data

F. Bouraoui and B. Grizzetti
European Commission, Joint Research Centre, Directorate D – Sustainable Resources, D2 – Water and Marine Resources Unit, Ispra (Va), Italy

ABSTRACT

The Nexus approach is often presented as an evolution and improvement of the existing integrated water resources management as it promotes a multi-centric approach integrating water, energy, and food policy objectives. Being a relatively new concept and being complex due to the interlinkages between water food and energy, the Nexus has no clear definition and no agreed conceptual framework for implementation. This chapter discusses how to move from concepts to implementation detailing some of the tools and approaches proper of the Nexus and the related data requirement. We discuss in detail qualitative and quantitative methodologies used to perform WEF Nexus assessments. We show that the availability of data and their integration from different policy sectors remain a strong limiting factor. We conclude highlighting relevant aspects to be considered when designing a WEF Nexus assessment.

Keywords: WEF Nexus, IWRM, data, tools

CS1.1 INTRODUCTION

The Nexus approach is often presented as an evolution and improvement of the existing integrated water resources management (IWRM). It has gained attention in the recent years as it is multi-centric and integrates water, energy, and food policy objectives (Benson *et al.*, 2015). The Nexus approach treats food, water, and energy as an interrelated system of systems (Eftelioglu *et al.*, 2017), and seeks to reduce antagonism and to assess trade-offs and synergies between these three interlinked pillars in view of a more sustainable development and use of natural resources.

By their interlinked nature, water, food, and energy require integrated and transdisciplinary approaches for addressing their nexus (Sønderberg & Larsen, 2016) across a large span of temporal and spatial scales. However, integration goes beyond the three sectors and should include social, political, and governance aspects. In addition, the Nexus approach should consider trade-offs, not only across sectors but also among different users of the same sectors (Sønderberg & Larsen, 2016).

In this context, data and appropriate methodologies are needed to inform on these complex interlinkages, in the present situation and also under different future scenarios, to help policy makers in decision making. This paper discusses how to move from concepts to implementation, detailing some of the tools, and approaches used to address the Nexus along with their data requirements.

CS1.2 TOOLS AND METHODOLOGIES FOR ASSESSING THE WATER–ENERGY–FOOD NEXUS

Being a relatively new concept and being complex due to the interlinkages between water, food, and energy (WEF), the Nexus has no clear definition and no agreed conceptual framework for implementation. The term Nexus has been used extensively, however, very little literature is available on concrete examples of Nexus studies.

An additional difficulty lies in the fact that the Nexus approach covers, in theory, a wide array of temporal and spatial scales ranging from local to regional, national and global, with all scales being related and impacting each other: global initiatives to combat climate change undoubtedly have an impact on local management as well as sectorial policies such as energy are usually designed at the national scale with strong implications at the local scale (e.g., hydropower and irrigation).

The existing frameworks usually propose a sequential approach to address the WEF Nexus and can be summarized as follows: (1) assessment of current status and trends; (2) assessment of the linkages to quantify trade-offs and synergies; (3) development of plausible scenarios and associated uncertainties, and identification of adaptation solutions; and (4) selection through consultation of the most appropriate interventions in terms of economic, social, environmental, and governance criteria. Additional steps might include implementation and monitoring and improvement (FAO, 2014; Bizikova *et al.*, 2013).

The existing tools used in WEF studies address one or many of the steps described above. Sønderberg and Larsen (2016), based on Granit *et al.* (2013), classify the tools used in Nexus assessments into four categories listed below in order of increasing data requirements:

- qualitative indicators-based methods;
- hydro-economic modelling;
- integrated WEF Nexus; and
- operational systems.

Differently, Endo (2015) used a simplified classification schemes limited to qualitative and quantitative assessments tools. Following this classification scheme, we discuss here the most common tools used in WEF Nexus assessments while describing the different methodologies, data requirements, and temporal and spatial scales of application (Figure CS1.1). Most of the tools in use are based on System Thinking (Reynolds & Holvell, 2010), which is an approach that helps understand the non-linear behaviour, linkages and interactions of complex systems, and provide information to support decision making.

CS1.2.1 Qualitative approaches

Qualitative approaches are used in the FAO Nexus Assessment Framework as a mean to layout the context of the analysis to understand the priorities, and the environmental, economic, and societal trade-offs. They often rely on describing and mapping the system through interaction with various stakeholders. The mapping consists of nodes, which represent the different concepts that describe the behaviour of the system, and arcs, which represent the relationship between any two nodes and the attributes which include the strength and direction of the relationship. Using such a cognitive model, one can simulate the effects of possible actions considering the perceived influences between the different elements of the system.

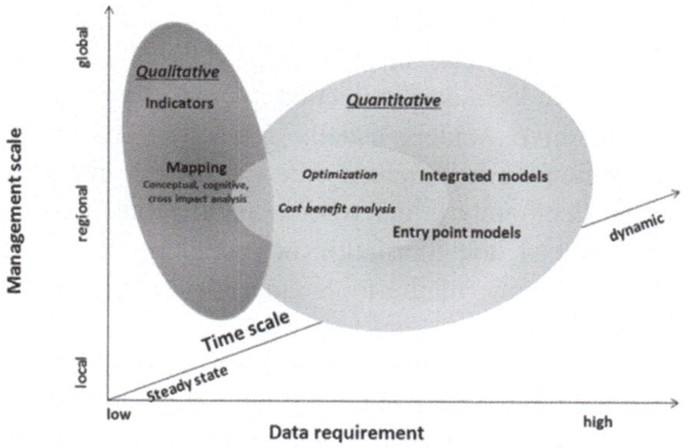

Figure CS1.1 Classification of tools used in WEF Nexus assessments.

Endo (2015) also used locally ontology engineering that is defined as formal representations of a set of concepts within a domain and the relationships between those concepts. The result can be a conceptual map that can be used to identify the problem to be addressed. Fuzzy cognitive mapping is based on the concept and causal relationships reasoning and is used to describe the functioning of complex systems characterized by large uncertainties such as those found in political science.

Questionnaire surveys have also been used to assess how local population's security is affected due to natural or social shocks. For example, Endo (2015) developed the questionnaire on the WEF system using the concepts of accessibility, utilization, and management. They found the questionnaire useful in deriving knowledge on the local population economic, livelihood, and food security under various WEF scenarios. A similar approach could be used at macro-region levels where questionnaires could be filled by experts panels, stakeholders etc. to derive a set of global indicators.

Cross impact analysis has also been used to analyse the dependencies and interdependencies among sectors in order to predict plausible scenarios of the future. This approach relies on expert panels that are asked to derive probabilities and conditional probabilities of occurrence (impact) of events. For instance, the Transboundary Waters Opportunity Analysis (Phillips *et al.*, 2008) was developed to promote a sustainable and equitable use of water resources, and identify opportunities for development opportunities. The analysis is matrix based and uses four key development opportunities, namely: (i) hydropower and power trading, (ii) primary production, (iii) urban and industrial development, and (iv) environment and ecosystem services as well as two categories of freshwater sources (surface and groundwater). The approach consists in identifying potential benefits at the basin level that are then analysed by all riparian countries to select win-win solutions.

Finally, Daher and Mohtar (2015) have proposed an approach based on indicators. They developed the WEF Nexus tool that combined resources indices (ratio of required resources and allowable capacity) including water, land, energy, carbon, financial into a sustainable indicator that should support decision makers in their choice of the most appropriate scenario.

CS1.2.2 Quantitative approaches

FAO, in its conceptual framework, adopts quantitative assessments for scenario analysis, for instance to evaluate the impact of alterative scenarios of intervention. Quantitative methods can also be used to assess actual status in order to identify priorities.

Quantitative approaches are usually more data intensive than qualitative approaches. However, data requirements are also controlled by the scale of application. Global issues require large-scale dataset, while local problems require intensive local data collection in order to adequately capture the local reality.

Most of the available quantitative models usually focus on one or two of the Nexus pillars (Miralles-Wilhelm, 2016), and often do not include external factors impacting the water, energy, and water sectors, such as land use, demography, or environmental protection making their use for policy option selection rather limited. Indeed, the translation of modelling studies in the Nexus context into policy decision is limited. A list of some of the tools currently in use for WEF Nexus assessments is given in FAO (2014) and IRENA (2015).

In our classification (Figure CS1.1), we distinguish as reported by IRENA (2015) between entry point models that assess the influence of one sector onto the other sectors, and fully integrated models that consider the bi-directional interactions between all sectors. Integrated models explicitly represent, with a wide range of complexity, the mechanisms and the processes and feedbacks involved in water, food, and energy cycles. The complexity stems not only from the description of the processes, but also from the spatial and temporal representation of the system boundaries (Bouraoui & Grizzetti, 2014). Integrated physical models are then often associated with an economic module and linked to optimization and cost–benefit analysis tools to evaluate the sustainability of alternative scenarios and also define efficient resource allocation strategies.

Few attempts have been made to address the WEF Nexus by integrating different pieces of software such as CLEWS (Howell *et al.*, 2013), which puts together the energy model (LEAP), the water model (WEAP) and the land-use model (AEZ). In CLEWS, interactions between the three modules are managed by exchanging data in an iterative manner. The model is semi-spatially explicit with simplified representations of the water cycles and crop growth. Other integrated approaches such as CGAM provide spatially aggregated quantifications of the WEF Nexus. The Pacific Northwest National Laboratory developed the PRIMA (Platform for Regional Integrated Modeling and Analysis) to simulate the complex interactions among climate, energy, water, and land at decision-relevant spatial scales. However, most of the applications so far have focused on the water–energy Nexus and water–food Nexus, and these studies usually remain in the scientific sphere and are not implemented at the policy- or decision-making level (Leck, 2015).

Efforts have also been made to link integrated physical model with optimization techniques for water allocation purposes or identification of the most appropriate water-saving measures (i.e., hydro-economic modelling). For instance, Udias (2016) linked the water resources model LISFLOOD with a multi-criteria optimization model to assess spatially explicit combinations of measures, which could help reduce the gap between water demand and water availability while taking into account ecological, water quality, flood risk, and economic aspects in the Danube River Basin. Pastori (2015) used a multi-objective optimization approach to identify optimum crop and land management patterns in different African countries. They provided trade-off alternatives that maximize crop production by choosing the adequate crop, fertilization, and irrigation management sequences while limiting the impact on the environment. Cost–benefit analysis and optimization techniques have also been used along with qualitative approaches to support stakeholder choices among different alternatives and also, for instance, in the selection of appropriate weights when combining different indicators.

CS1.2.3 Data constraints

Several tools as described before have been developed and used in the context of the WEF Nexus. However, our understanding and representation of the interactions and trade-offs are often limited by data availability, collection and management (Eftelioglu, 2017). Indeed, data are the foundation for validating and improving the model (Wolfe *et al.*, 2016). Eftelioglu (2017) lists limited observability and accounting of the interactions between the resources, computational requirements, and lack of organized and harmonized data-sharing protocols as severe limitations to our in-depth understanding of the WEF Nexus. Similarly, IRENA (2015) points out the difficulty in gathering relevant data due to the cross-sectorial and multi-scale nature of the Nexus.

Most of the data available are usually focused on one sector, with however a few attempts to capture some of the inter-connections with other sectors: water use in irrigation for instance. However, most of the data fail to encompass a comprehensive overview of the interlinkages. This is where a large effort is still to be undertaken. In addition, data on water, energy, and food sectors are also collected under different institutions or departments according to the priorities and objectives of the respective sectorial policies, and often at different spatial and temporal scales that hamper an easy harmonization. Furthermore, a rigorous assessment of the WEF Nexus requires collecting data in the long term so as to use these time series to understand if the data (including measurements, model predictions, trends, and projections) alongside its derived indicators and results change coherently with the relative management decisions and interventions so they can be used routinely for assessments and evaluation.

It is also of key importance to integrate uncertainty when collecting data. Indeed, data are collected to represent the past or the near present. Extrapolating data to predict trends and trajectories requires the integration of uncertainties in projections. Uncertainty is also embedded in the measured data used to represent the baseline or past assessments, and it is also present when resulting from model predictions (climate change, for instance). All these uncertainties need to be recognized, quantified, and incorporated in the assessments to increase the reliability and acceptability of the Nexus assessments.

Data acquisition and accessibility is recognized as the limiting factor to a successful implementation of the WEF assessment, and convergence thinking has been proposed to overcome discipline boundaries and integrate knowledge from the physical, biological, social, economic, and mathematical sciences (Wolfe *et al.*, 2016). Comparability of data across sectors is challenging due to the lack of harmonized protocols to gather, share, and interpret the information. As part of its Nexus Platform, UNU-FLORES advocates a harmonization of data collection protocols and access to complete data, and calls for a unified monitoring framework (UNU-FLORES, 2015). To facilitate the exchange of data and information across disciplines, an integrated data management framework is needed. This framework should allow assimilation of data from different spatial representation starting from a commonly agreed metadata. An additional limitation to data sharing is linked to proprietary and confidentiality issues (McCarl *et al.*, 2017). Data gaps need to be determined and models are a natural alternative for identifying additional data requirements. Prospective analysis will also be needed to generate data about future technological developments and how these can be assimilated by the current tools.

CS1.3 CONCLUSIONS AND RECOMMENDATIONS

Moving from the concept to the implementation of the WEF Nexus in river basin management is challenging. A holistic conceptual framework is needed to describe the interlinkages between water, energy, and food, and appropriate quantification tools that consider the multiple spatial scales involved in the scope of the impacts of the actions, from local to regional to global (and transboundary for water issues). This is in part addressed by the qualitative and quantitative approaches discussed here. The availability of data and their integration from different policy sectors remains a strong limiting factor in quantitative assessments of WEF Nexus, especially in countries lacking sufficient investments on data infrastructures. Important advances in this regard might be represented by satellite data, citizens' participation in data collection or automated sensors in monitoring. Importantly, the WEF Nexus should include stakeholders in the analysis of interlinkages and impacts. This involves considering interdisciplinary and transdisciplinary analysis, that should be reflected in the data, tools and expertise deployed in a WEF Nexus assessment. Impacts on the environment at the local, regional, and the global scales should be embedded in WEF Nexus assessments as well as issues related to inequalities in the distribution of impacts. The ecosystem services approach (de Groot, 2010) could be useful in this regard to understand the effects of policy actions on different perspectives, including the analysis of beneficiaries, and also to help translate the results of WEF Nexus assessments to policy makers and people.

Based on these considerations, we can formulate the recommendation to consider the following aspects when designing a WEF Nexus assessment and choosing the methodology for quantification:

- data availability,
- temporal scale and the spatial scale of impacts,
- stakeholder involvement,
- impacts on ecosystems at the local, regional, and global scale,
- inequalities in the distribution of impacts on people's well-being,
- interdisciplinary attributes of experts and tools, and
- communication needs in translating results to policy makers and people.

By integrating objectives from different sectorial policies, assessments based on Nexus approach and methodologies result particularly appropriate for achieving the SDGs taking into account the complex interlinkages between the different goals.

REFERENCES

Benson D., Gain A. K. and Rouillard J. J. (2015). Water governance in a comparative perspective: from IWRM to a 'nexus' approach? *Water Alternatives*, **8**, 756–773.

Bizikova L, Roy D., Swanson D, Venema H. D. and McCandles M. (2013). The Water–Energy–Food Security Nexus: towards a practical planning and decision-support framework for landscape investment and risk management. IISD report, February 2013.

Bouraoui F. and Grizzetti B. (2014). Modelling mitigation options to reduce diffuse nitrogen water pollution from agriculture. *Science of the Total Environment*, **468–469**, 1267–1277.

Daher B. T. and Mohtar R. H. (2015). Water–energy–food (WEF) Nexus Tool 2.0: guiding integrative resource planning and decision-making, *Water International*, **40**(5–6), 748–771.

de Groot R. S., Alkemade R., Braat L., Hein L. and Willemen L. (2010). Challenges in integrating the concept of ecosystem services and values in landscape planning, management and decision making. *Ecological Complexity*, **7**, 260–272.

Eftelioglu E., Jiang Z., Tang X. and Shekhar S. (2017). The nexus of food, energy, and water resources: visions and challenges in spatial computing. In: Advances in Geocomputation. Advances in Geographic Information Science, D. Griffith, Y. Chun and D. Dean (eds.) Springer.

Endo A., Tsurita I., Burnett K. and Orencio P. M. (2015). A review of the current state of research on the water, energy, and food nexus. *Journal of Hydrology: Regional Studies*, **11**, 20–30.

FAO (2014). Walking the Nexus talk: Assessing the Water-Energy-Food Nexus in the context of the sustainable energy for all initiative environment and natural resources. Working Paper No. 58, FAO, Rome.

Granit J., Fogde M., Hoff H., Joyce J., Karlberg L., Kuylenstierna J. and Rosemarin A. (2013). Unpacking the water-energy-food nexus: tools for assessment and cooperation along a continuum. In Cooperation for a Water Wise World—Partnerships for Sustainable Development; Report Nr. **32**, 45–50. Stockholm International Water Institute: Stockholm, Sweden.

Howell M., Hermann S. and Welsch M. *et al.*, 2013. Integrated analysis of climate change, land-use, energy and water strategies. *Nature Climate Change*, **3**, 621–626.

IRENA (2015). Renewable Energy in the Water, Energy, and Food Nexus, International Renewable Energy Publication, Abu Dhabi.

Leck H., Conway D., Bradshaw M. and Rees J. (2015). Tracing the water–energy–food nexus: description. *Theory and Practice. Geography Compass*, **9**, 445–460.

McCarl B. A., Yang Y., Schwabe K. and Engel B. A. (2017). Modeling and data for WEF nexus analysis: a review of issues. FEW nexus workshop on integrated science, Engineering, and Policy: A Multi Stakeholder Dialogue January 26–27, 2017, College Station Texas.

Miralles-Wilhelm F. (2016). Development and application of integrative modeling tools in support of food-energy-water nexus planning—a research agenda. *Journal of Environmental Studies and Sciences*, **6**, 3.

Pastori M., Udías A., Bouraoui F., Aloe A. and Bidoglio G. (2015). Multi-objective optimization for improved agricultural water and nitrogen management in selected regions of Africa. *International Series in Operations Research and Management Science*, **224**, 241–258.

Phillips D. J. H., Allan J. A., Claassen M., Granit J., Jägerskog A., Kistin E., Patrick M. and Turton A. (2008). The TWO Analysis: Introducing a Methodology for the Transboundary Waters Opportunity Analysis. Report Nr. 23. SIWI, Stockholm.

Reynolds M. and Holvell S. (2010). Systems Approaches to Managing Change: A Practical Guide. Springer.

Sønderberg P. L. and Larsen H. H. (eds) (2016), DTU International Energy Report 2016: The Energy-Water-Food Nexus—from local to global aspects. Technical University of Denmark (DTU).

Udias A., Gentile A., Burek P., De Roo A., Bouraoui F., Vandecasteele I., Lavalle C. and Bidoglio G. (2016). Multi-criteria framework to assess large scale water resources policy measures. *Water*, **8**, 370.

UNU-FLROES (2015). Nexus Observatory Platform. https://nexusobservatory.flores.unu.edu/ (last access July 2021).

Wolfe M. L., Ting K. C., Scott N., Sharpley A., Jones J. W. and Verma L. (2016). Engineering solutions for food-energy-water systems: it is more than engineering. *Journal of Environmental Studies and Sciences*, **6**, 172–182.

Tools for water security contributing to WEFE Nexus

A. Mishra[1], Y. Filali[1], A. Makarigakis[1], K. Verbist[1,2], A. Amani[1] and B. Jimenez Cisneros[1]

[1]UNESCO Water Sciences Division, Paris, France
[2]UNESCO Regional Office for Southern Africa, Zimbabwe

ABSTRACT

The UNESCO Intergovernmental Hydrological Programme (UNESCO-IHP) contributes to dealing with complex interlinkages, and rapid environmental and demographical pressures through holistic, transdisciplinary and environmentally sound approaches to water resources management in line with international agendas. The need for reliable information to successfully deal with complexity in managing water has led to the development and promotion of a number of tools and methodologies. These combine various types of data that harness information and communications technology (ICT) as well as modelling to address water security challenges and tentatively the WEFE Nexus. They include, among others, the deployment of hydro-climate monitoring systems in Latin America and Africa, capacity building for flood warning and management in Pakistan and Afghanistan, and climate risk management in urban areas. To address these challenges, the UNESCO-IHP launched its Water Information Network System (IHP-WINS) in January 2017. IHP-WINS is an open-access, data- and knowledge-sharing platform for water-related issues at all levels, and is freely available to UNESCO Member States and all other water stakeholders. UNESCO-IHP promotes open access to innovative, free and open-source software and applications for water management, particularly for partners in developing countries. The different examples presented in the following section show how integrating various data sources, ICT and hydro-climate modelling can provide tools for pursuing water security and efficient WEFE Nexus opportunities.

Keywords: WEFE Nexus, hydro-climate, drought vulnerability, flood management, water information network, water free-open-access software

CS2.1 BACKGROUND

In many regions of the world, changes in precipitation and melting of snow and ice are altering hydrological systems and consequently affecting water resources in terms of quantity and quality (IPCC, 2014). Despite the fundamental value of freshwater to public health, agriculture, industry, prosperity, and security, challenges related to water scarcity, pollution, poor sanitation, and water-related disasters confront billions of people worldwide. Almost half of the world's population will be living in areas of high water stress by 2030.

Currently and globally, approximately 80% of the world's population already suffer serious threats to its water security, as measured by indicators covering the different criteria of water availability, accessibility, safety and quality and management, and therefore including water quantity, water demand per need or pollution. Still nearly a billion people do not have proper access to safe water and 2.5 billion to safe sanitation (UNESCO/WWDR, 2017). From those enjoying these services, more than 60% do not have a proper and reliable service due to intermittent supply. About 500 million people live in areas where water consumption exceeds the locally renewable water resources by a factor of 2 (Mekonnen & Hoekstra, 2016).

Water-related risks will further increase as a result of growing climate instability, population growth, and forced migration; all of which will put additional pressure on the water resources of both the host and neighbouring countries. One of the greatest challenges for the hydrological community is to identify appropriate and timely adaptation measures in this continuously changing environment, and establish sectorial interlinkages towards achieving SDGs and targets under the Paris Agreement and the Sendai Framework for Disaster Risk Reduction.

UNESCO's Intergovernmental Hydrological Programme (IHP) has long been implementing projects related to knowledge generation and capacity building for water security under global changes, including climate variability and change, population growth, urbanization and economic development. The IHP is in particular studying the impacts of global change on water resource systems, including ways to enhance resilience to climate-related disasters, and address both urban and rural water needs. To successfully deal with complexity in water management, tools and methodologies combining various types of data that harness information and communications technology (ICT) as well as modelling were developed, and are being promoted and implemented. These include the deployment of a hydro-climate monitoring system in Latin America and Africa, capacity building for flood warning and management in Pakistan and Afghanistan, climate risk management in urban areas, and the Water Information Network System.

CS2.2 HYDRO-CLIMATE MONITORING FOR IMPROVED WATER MANAGEMENT AND FOOD SECURITY IN AFRICA AND LATIN AMERICA

Through its initiatives related to hydrological changes, UNESCO-IHP supports countries in identifying and addressing their information gaps and needs when it comes to managing the risks resulting from the changing hydrological and climatic impacts. It does that by strengthening global, regional, and local capacities and by providing access to data and policy recommendations for more integrated hydro-climate risk management. To that effect, UNESCO-IHP supported the development of an integrated flood and drought monitoring and forecasting system for Africa and Latin America (Verbist *et al.*, 2016). The system (http://stream.princeton.edu/) developed by the University of Princeton in the United States combines remote-sensing data on precipitation, vegetation, and atmospheric analysis with macro-hydrological modelling through the use of a variable infiltration capacity (VIC) land-surface hydrological model (Liang *et al.*, 1994). The system tracks hydrological conditions including extremes (e.g. floods and droughts) in near real time and allows medium-term and seasonal forecast. It therefore provides monitoring capabilities for meteorological, hydrological, and agricultural drought and flood conditions which is particularly useful in developing regions where institutional capacity for monitoring and early warning is generally lacking and access to information and technology prevents the development of such systems locally. In addition, the system has the advantage of providing a standardized format for any of the components of the water balance, thus enabling a comprehensive analysis of drought and flood hazards at local, national, and regional level. In essence, the system provides information on precipitation, temperature, radiation and wind speed, drought indicators (i.e. Standardized Precipitation Index – SPI, soil moisture, Normalized Difference Vegetation Index – NDVI, evapotranspiration) and flood indicators (i.e. surface runoff and streamflow). The information can be obtained either spatially or for point locations, for specific dates, months, or annual timescales, and is compared with the normally expected conditions or percentiles. The system was successfully deployed in Western, Eastern and Southern Africa combined with training of experts and is used as a complementary information system by regional institutions to monitor agro-hydro-meteorological conditions particularly during the rainy seasons.

Similarly, UNESCO-IHP has collaborated with the Centre for Hydrometeorology and Remote Sensing (CHRS) at the University of California, Irvine, on the development of tools to provide near real-time global satellite precipitation estimates at high spatial and temporal resolutions, including the Precipitation Estimation from Remotely Sensed Information using Artificial Neural Networks-Cloud Classification System (PERSIANN-CCS) (Hsu *et al.*, 2010). This specific system provides essential information for emergency planning and management of hydrological risks, such as floods, droughts, and other extreme weather events. For example, the Namibia Hydrological Services (NHS) uses this system to prepare daily bulletins with information on flood and drought conditions for local communities. The system is now available through the iRain mobile application, specially designed to facilitate people's involvement in collecting local data for global precipitation monitoring (http://en.unesco.org/news/irain-new-mobile-app-promote-citizen-science-and-support-water-management). iRain allows users to visualize real-time global satellite precipitation observations, track extreme precipitation events worldwide, and report local rainfall information using a crowd-sourcing functionality to supplement these data and also provide ground information which can improve remote sensing precipitation estimations.

CS2.2.1 Key findings

In developing low-income regions such as Latin America and Africa, data needed for implementation of drought monitoring systems are scattered over multiple agencies that are dependent on different ministries. This requires collaboration across ministries through a multi-sectorial approach, which often cannot be effectively implemented without direct support from high-level policy makers. Monitoring and early warning systems require combining data sources from national weather and hydrological services, agricultural extension services and public databases as well as data streams from international partners providing remote sensing datasets to fill data gaps and global/regional weather and climate model outputs. The challenge of this approach is that it requires technological solutions that allow the integration of multiple data sources with different temporal and spatial resolutions. An additional challenge is that these sources often have a complex data structure and data exchange formats that need to be treated to allow their integration in a seamlessly working system.

CS2.3 DROUGHT VULNERABILITY ATLAS AND OBSERVATORY IN CHILE AND PERU

In close collaboration with the Chilean Ministry of Agriculture, the Food and Agriculture Organization (FAO) and the International Research Institute for Climate and Society (IRI), the Chilean Agroclimatic Observatory system (www.climatedatalibrary.cl/UNEA/maproom/) was launched in June 2013. A similar observation system was developed in collaboration with the Autoridad Nacional del Agua (ANA) in Peru in 2014 (http://ons.snirh.gob.pe/Peru/maproom/). These systems allow for the creation of integrated indices, taking into account a number of different drought indicators. The systems build upon the Climate Data Library (CDL), a tool that allows the collection of all raw databases of national and international institutions relevant to drought monitoring (Del Corral *et al.*, 2012). Data of numerous formats can be added, with which additional indicators can be calculated using advanced arithmetic or geo-statistical functions. In order to provide effective decision-support tools, a user-friendly interface was built on top of the CDL, called the "maproom", which holds relevant drought indices on meteorological, hydrological, and agricultural drought, and combines information from national and international datasets.

Support was also provided by UNESCO for the development of an (agricultural) drought vulnerability atlas for Chile, focussing on vulnerability of rural communities, using indicators related to agricultural production and rural poverty (i.e., sensitivity, exposure, and adaptation capacity) and covering socio-economic, biophysical, and institutional aspects. In order to provide an objective and actionable drought index, information from meteorological, hydrological, and agricultural drought indicators were integrated into a Combined Drought Index (CDI) for Chile, which was based upon the CDI developed for the European Drought Observatory (Sepulcro-Canto *et al.*, 2012). Three standardized drought alert phases have been defined by their drought intensities. The standardizing of these alert phases for the whole of the country allows for an implementation of a more objective drought management response.

CS2.3.1 Key findings

The development of a national drought observatory in Chile and Peru provided an example of how to build a system to support integrated risk management decisions related to the WEFE nexus. By gathering all relevant national actors around the table to share data and information, the observatories have seen a continuous expansion from a purely drought focus to include indicators of agricultural crop failure risk and hydropower generation evaluation in real time. It has also driven the need to identify socio-economical drivers of drought vulnerability, which often include environmental aspects such as water quality. As such, the observatories are a key contributor to effective (drought) risk management policies related to the WEFE Nexus and also highlighted in the Sendai Framework and the SDGs.

CS2.4 STRENGTHENED CAPACITY FOR FLOOD WARNING AND MANAGEMENT IN PAKISTAN

Following the devastating 2010 floods in Pakistan, UNESCO successfully implemented, in partnership with Japan International Cooperation Agency (JICA), a major project enhancing capacity on flood warning and management. The objectives of this project were to reduce human and socio-economic impacts of floods in Pakistan, while improving the social, economic, and possible ecological benefits from flood activities and fostering safer human settlements near floodplains. A new flood forecasting and early warning system for the Indus River was developed, enabling the use of satellite-based rainfall observation data to complement the limited capacity of the ground observation system in Pakistan. A new model was developed for the Lower Indus area affected by the 2010 floods that has enabled Pakistani agencies to conduct flood inundation analysis and develop hazard maps. Following expansion of the model to the Eastern Rivers (Jhelum, Chenab, Ravi, and Sutlej), a prototype of the current unified Indus Integrated Flood Management System (Indus-IFAS) (http://www.icharm.pwri.go.jp/research/ifas/) covering the whole Indus River Basin was delivered. The technical capacity of more than 1000 experts from various agencies in Pakistan was enhanced for flood management forecasting, early warning and flood hazard mapping and analysis. Strengthened cooperation with other Indus river basin countries (i.e., Afghanistan, India, and China) for transboundary flood management and data sharing was also promoted by UNESCO.

CS2.4.1 Key findings

The establishment of a multi-stakeholder platform comprising partners dealing with flood management was critical. Training programmes on flood management for officials and experts from Afghanistan were mainly provided by Pakistani officials and experts. This was the first case of this type of exchange in the region, and has helped strengthen cooperation between experts of the two countries.

CS2.5 CLIMATE RISK MANAGEMENT WITHIN URBAN AREAS

According to the UN, 54% of the world's population lives in urban areas, a proportion that is expected to increase to 66% by 2050. It is projected that 2.5 billion people will be added to urban populations by 2050, 90% of which will be in Asia and Africa (UN, 2014). In 1990, there were 10 "megacities" (cities with 10 million inhabitants or more), which were home to 153 million people or slightly less than 7% of the global urban population at that time. As of 2014, there are 28 megacities worldwide; home to 453 million people or about 12% of the world's urban dwellers and by 2030, the world is projected to have up to 41 megacities (United Nations).

Megacities exert significant pressure to water resources located well outside their spatial coverage, covering quite often more than the river basin they are located in (e.g., Mexico City). This requires substantial amounts of energy. The rapid accumulation of people in these metropoles will translate into furthering the challenge of providing water and sanitation-related services from both a water resources and an energy point of view.

Smart water management employ ICT technology to enable urban water systems to be designed, controlled, and maintained in a way that allows optimization of water quantity, water quality, and the water energy footprint. By employing various sensors (pressure, element related, etc.) and automated systems (SCADA etc.), leaks can be identified rapidly minimizing the loss of the valuable resource and the quality of water can be guaranteed to the last client at the end of the water supply system.

UNESCO-IHP has been spearheading work on Smart Water System technologies in efforts to contribute tangible solutions to the WEFE Nexus, including raising awareness around these, training experts, and disseminating examples from pilots that may be replicated in other megacities. Through its work on 'Urban Waters', UNESCO has developed knowledge materials (see the Urban Water Series: https://en.unesco.org/uwmp/resources#Urban) that address the present challenges faced by cities worldwide. Through the publication of a monograph on water in 15 emblematic megacities, UNESCO has shared best practices to adapt to climate change and reduce the water and energy footprint in cities globally.

UNESCO is also working on the establishment of a collaborative platform to support megacities on adapting to or mitigating the effects of climate change related to water. In addition to learning from each other's experience and exchanging best practices, participating megacities will need to partner with the right technical, academic, and financial institutions, and design and implement their individual responses to the challenge. The platform (www.eaumega.org) will be using IHP-WINS as a knowledge-sharing tool for its implementation.

CS2.5.1 Key findings

Technologies and systems for smart water management need to be identified and made available through wide dissemination among cities' water utilities, particularly in developing countries. It is important to look at the interlinkages of energy and water footprints in both water supply and energy production and at relevant policies to develop that can promote these smart technologies and systems to reduce the footprints. It is important also to promote establishment of platforms for knowledge production, exchange, and dissemination as demonstrated by the next example.

CS2.6 WATER INFORMATION NETWORK AND PROMOTION OF WATER-FREE OPEN-ACCESS SOFTWARE

IHP's Water Information Network System (IHP-WINS) provides an open-access, knowledge-sharing, online platform, which was launched in January 2017 (UNESCO, 2017a). The platform is made available for free by UNESCO-IHP to UNESCO Member States, all other water stakeholders, partners, and individuals. It aims at facilitating access to knowledge, while users are encouraged to share data. As a result, the platform is continuously enriched with new data and information coming from various sources. In particular, IHP-WINS allows for sharing and access to water-related GIS data at various scales, which users can combine to create maps tailored to their needs (Figure CS2.1). By superimposing information on the spatial extent of transboundary aquifers and groundwater pollution risk, transboundary resources potentially at risk can quickly be identified. From this approach, areas where cooperation between states for water management should be encouraged can be developed. Transparency and respect of authorship is guaranteed and all information provided benefits from metadata in a standardized format. The latter format embeds a digital object identifier (DOI), which allows for an accurate identification and crediting of any contribution. The platform also aims at stimulating inter-disciplinary collaboration, professional networking and mentoring through working groups where users can exchange and provide feedback on their's and other's ongoing work. By gathering global and inclusive knowledge on water, and by facilitating interdisciplinary collaboration, the overall aim of IHP-WINS is to support UNESCO Members States and other stakeholders involved in water resources management. Additionally, due to its open-access nature, the platform also contributes to closing the gap between North and South, urban and rural areas and high- and low-income regions in terms of access to knowledge. Finally, the initiative contributes to the follow-up by the UNESCO Members States and the United Nations custodian agencies on the monitoring and implementation of the targets of Sustainable Development Goal 6 (SDG 6) and those of other water-related goals.

Continuing work on the reduction of the digital divide between developed and developing countries, UNESCO-IHP launched the HOPE Initiative in June 2013. This Hydro-Free and Open-source Platform of Experts aims to promote the use of Free and Open-Source Software (FOSS) and applications (UNESCO, 2017b). It provides a new approach to research that is more integrative, international, and solutions-oriented. HOPE links high-quality focused scientific research to new policy-relevant interdisciplinary efforts for global sustainability based on the use of FOSS. In partnership with 18 universities, centres, and other organizations, UNESCO is also collaborating on FREEWAT (FREE and open-source tools for WATer resource management project), that is promoting an innovative participatory approach gathering technical staff and relevant stakeholders, including policy and decision makers, to jointly design scenarios for the proper application of conjunctive water policies (FREEWAT, 2017).

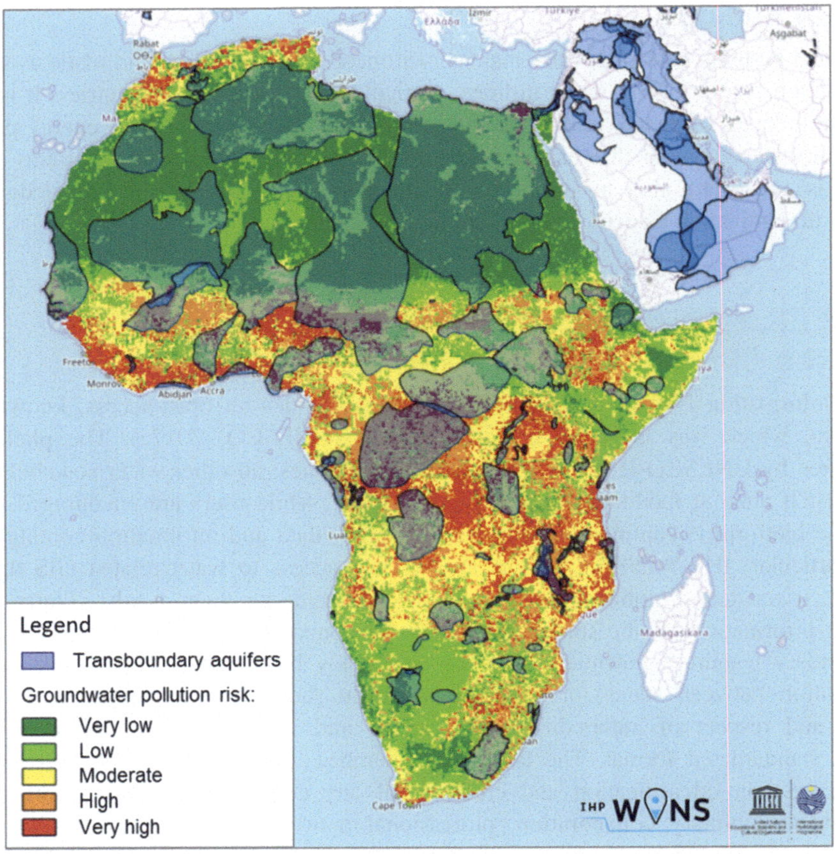

Figure CS2.1 Transboundary aquifers and groundwater pollution risk in Africa (IGRAC & UNESCO-IHP 2015; Ouedraogo *et al.*, 2016). © IHP-WINS.

CS2.6.1 Key findings

Accurate information on the trends of countries' water resources is required as a basis for economic and social development, and for maintaining environmental quality. In every sector of economic activity, planning, development, and operation require water-related information. With many competing uses and a finite amount of water, water resources need to be managed as effectively as possible, allowing for enough water, of sufficient quality, for everyone including the environment. To make sound decisions, decision-makers and other stakeholders rely increasingly on reliable, accessible data, and free water information systems.

CS2.7 DISCUSSION, CONCLUSIONS, AND RECOMMENDATIONS

Evidence based, inclusive decision making and a smarter way to use and share information is greatly needed to address the complex interlinkages within the WEFE Nexus and the SDGs. This requires integration of various sources and types of data and modelling to provide sound, actionable scientific information. Cooperation among stakeholders and decision makers from various sectors is also necessary.

The examples presented in this paper highlight the critical contribution of filling the data and knowledge gap to achieve a better understanding of the interlinkages that are indispensable for effective decision making. The examples also highlight the importance of ICT, smart water technologies, and modelling in helping to address the need for filling the data gaps and data integration from various sources and types as well as the critical importance of cooperation between different sectors. The role of water information network systems such as WINS allowing for the integration of various data layers, and sharing and exchanging of data and information, is paramount. It is also important to set up and actively use a multi-stakeholder platform to facilitate dialogue and inclusiveness in order to better address issues and themes such as the WEFE Nexus and SDG implementation challenges. Open-source and free ICT-related tools are recommended particularly for developing countries. Finally, building the capacity of both decision makers and technical agencies in charge of collecting, processing, and managing data, by equipping them with relevant and easy access tools, is indispensable.

REFERENCES

Del Corral J., Blumenthal M., Mantilla G., Ceccato P., Connor S. and Thomson M. (2012). Climate information for public health: the role of the IRI climate data library in an integrated knowledge system. *Geospatial Health*, **6**(3), 15.

FREEWAT (2017). Free and open source software tools for water resource management, from http://www.freewat.eu/ (Retrieved 28 October 2017).

Hsu K., Behrangi A., Iman B. and Sorooshian S. (2010). Extreme precipitation estimation using satellite-based PERSIANN-CCS algorithm. In: Satellite Rainfall Applications for Surface Hydrology, M. Gebremichael and F. Hossain (eds.), Springer, pp. 49–67.

IGRAC (International Groundwater Resources Assessment Centre) and UNESCO-IHP (UNESCO International Hydrological Programme), (2015). Transboundary Aquifers of the World [map]. Edition 2015. Scale 1:50 000 000. Delft, Netherlands: IGRAC.

IPCC (2014). Fifth Assessment Report on climate change.

Liang X., Lettenmaier D. P., Wood E. F. and Burges S. J. (1994). A simple hydrologically based model of land surface water and energy fluxes for general circulation models. *Journal of Geophysical Research*, **99**(D17), 14415–14428.

Mekonnen M. M. and Hoekstra A. Y. (2016). Four billon people facing severe water scarcity. *Science Advances*, **12**(2).

Ouedraogo I., Defourny P. and Vanclooster M. (2016). Mapping the groundwater vulnerability for pollution at the pan African scale, *Science of the Total Environment*, **544**, 939–953. https://doi.org/10.1016/j.scitotenv.2015.11.135

Sepulcre-Canto G., Horion S., Singleton A., Carrao H. and Vogt J. (2012). Development of a combined drought indicator to detect agricultural drought in Europe. *Natural Hazards and Earth System Sciences*, **12**, 3519–3531.

UN (2014). Department of Economic and Social Affairs, Population Division, World Urbanization Prospects: The 2014 Revision, Highlights (ST/ESA/SER.A/352

UNESCO (2017a). Welcome to the water information network system by the international hydrological programme of UNESCO, from http://ihp-wins.unesco.org/ (Retrieved 28 October 2017).

UNESCO (2017b). HOPE initiative: Hydro Free and Open-source Platform of Experts, from https://en.unesco.org/hope (Retrieved 28 October 2017).

UNESCO/World Water Development Report: (2017): Wastewater: A resource.

Verbist K., Amani A., Mishra A. and Cisneros B. J. (2016). Strengthening drought risk management and policy: UNESCO international hydrological programme's case studies from Africa and Latin America and the Caribbean. *Water Policy*, **18** (S2), 245–261, doi:10.2166/wp.2016.223.

The water and energy nexus in Africa

I. Hidalgo-Gonzalez[1], R. Gonzalez-Sanchez[2], P. Marcos-Garcia[2] and C. Carmona-Moreno[2]

[1]European Commission, Joint Research Centre (JRC), Petten, Netherlands
[2]European Commission, Joint Research Centre (JRC), Ispra, Italy

ABSTRACT

This contribution explores the complex interactions between water and the energy industry in Africa. Energy is needed for extraction, treatment and distribution of drinking water, wastewater treatment, and desalination. On the other hand, water is needed for electricity production, fossil-fuel extraction, transport and processing or for irrigating energy crops. Therefore, the availability and temperature of water resources act as constraining factors in the operation of power plants, which need water for cooling and for hydropower generation. In this regard, electricity and water demands are expected to grow significantly in Africa in the next decades. The African energy systems are small sized, poorly electrified, reliant on oil for power generation and poorly interconnected. As hydropower is the dominant renewable energy source in the continent, climate variability has a strong impact on the energy mix, operational costs, CO_2 emissions and water consumption for energy generation. Africa has a large untapped hydropower potential, however, new developments should be carefully assessed in regions where water scarcity is already an issue. Hydropower is very water-intensive due to evaporation losses in reservoirs, thus the substitution of fossil fuels by non-hydro renewable energies (such as wind or solar) could reduce significantly water use while helping to meet the increasing energy needs of the continent.

Keywords: WEFE Nexus, water, energy, African power pools

CS3.1 INTRODUCTION

The water–energy nexus is particularly challenging in Africa. The combined effect of economic and demographic growth is expected to soar electricity and water needs up to 700% and 500%, respectively, in 2050 (when compared to 2012) (World Bank, 2013). Energy and water demands cannot be fully decoupled. The growing energy demand will add pressure on water resources in certain areas (which are already characterized by water scarcity), competing directly with other sectors (e.g., agriculture, urban supply) and triggering potential conflicts among water users.

Water availability and water temperature have an effect on cooling systems which can constrain power plants' operation. Currently, hydropower is the dominant renewable energy source in most African energy systems, supplying up to 51% of electricity in sub-Saharan Africa in 2018 (excluding South Africa) (IEA, 2019). In this context, energy supply disruptions due to droughts frequently lead to negative economic and health aftermaths in African countries, affecting the energy mix, operational costs, CO_2 emissions, and water consumption for energy generation (Hidalgo Gonzalez *et al.*, 2021).

African energy issues have received increased attention on the European policy agenda and this focus has been further elevated with the recent communications on the European Green Deal (EC, 2019), which stresses that "climate and environmental issues should be key strands in relations between the two continents" and the communication on a comprehensive strategy with Africa (EC, 2020). Among the ongoing initiatives, the Water–Energy–Food–Ecosystems (WEFE) project of the Joint Research Centre (JRC) aims to support the design and implementation of cross-sectoral policies, to improve the resilience of water-using sectors and the preservation and sustainability of freshwater resources. Concretely, the analysis of the water–power nexus in Africa has been addressed within the context of the WEFE project, through the development of a specific modelling framework able to quantify the economic impacts, emissions, water withdrawals and consumption, and the detailed operation of the power system under current and future scenarios.

CS3.2 THE AFRICAN POWER POOLS

Currently, five power pools are established in Africa: West, Southern, North, Eastern and Central African Power Pool, respectively. Their main goal is to coordinate power system planning and operation across their members, which is reflected in growing interconnection levels and a gradual implementation of market-based integrated approaches. Future interconnections between power pools are also planned.

CS3.3 WESTERN AFRICAN POWER POOL

The West African Power Pool (WAPP) is characterized by its high vulnerability to climate change, and it is already experiencing impacts on its food, water, and energy security which could be further challenged by a rise of the demand associated with economic and population growth. Although the region is rich in water resources (approximately 27% of Africa's internal renewable water resources), it suffers from chronic water deficits due to the uneven distribution of rainfall and flows in time and space, insufficient knowledge about water resources, low exploitation of potential resources, and poor resource management. Energy resources are also plentiful but unenvenly distributed, and the renewable energy potential is underused. Electrification rates are low and there is a high dependence on biomass. The power generation mix has a significant share of gas and oil power plants and the interconnections between countries are very limited.

The operation of the WAPP's power system in the future will depend heavily on the availability of water resources, which is nevertheless outside the control of policy planning. This translates into a high volatility of the system cost, which can be partially mitigated by the addition of thermal capacities foreseen in the WAPP's master plan. However, that would lead to increased emissions and a higher electricity bill. Hence, future policy scenarios should explore technology portfolios which could achieve low volatility, low cost and low emissions simultaneously (De Felice *et al.*, 2019).

CS3.4 SOUTHERN AFRICAN POWER POOL

The Southern African Power Pool (SAPP) is the most advanced among the African power pools in terms of market development. Its member countries are quite heterogeneous as regards population size and economic development. Power generation in the SAPP relies largely on coal and water from the Congo, Orange, and Zambezi rivers and their tributaries. Both energy sources are water intensive, making regional electricity generation prone to water-related impacts that, in conjunction with other factors, can cause extended periods of electricity supply interruptions with high economic costs (in Malawi, South Africa and Tanzania have been estimated at 5–7% of the gross domestic product (GDP)), which could be even worse in the future. In addition, the planned developments will increase the concentration of the SAPP's hydropower capacity in the Zambezi basin, from about 70% to 85% in 2030. This growing accumulation of hydropower in a single basin could escalate the future water-related risks in the SAPP (Conway *et al.*, 2017).

The analysis of the water–power nexus in the SAPP shows that the discharge variability has caused electricity supply interruptions in recent years, mostly in those SAPP countries not yet interconnected, namely Angola, Tanzania, and Malawi (where the levels of unserved energy could reach up to 25% of the yearly country demand). The results also point out that a higher availability of water can substantially alleviate the negative economic consequences of unserved electricity in the SAPP (both on electricity price levels and hampered economic activity). Besides, it would be necessary to address the future impacts of climate change and the occurrence of extreme events (in South Africa, floods already reduced the operation of the coal fleet with impacts on several SAPP countries). In this regard, the expansion of interconnections could increase the resilience against electricity supply interruptions, significantly reducing and smoothing electricity prices and the unserved electricity levels. Better interconnection of the SAPP countries could reduce the system costs by 20%. Therefore, emphasis should be placed on grid expansion policies, as they are the only ones which can be directly controlled through policy decisions (Busch *et al.*, 2020).

CS3.5 THE NORTH, EASTERN, AND CENTRAL AFRICAN POWER POOLS

The renewable and fossil potentials vary significantly between the three remaining power pools. The North Africa Power Pool (NAPP) is mostly dominated by fossil fuels and has higher electrification rates, while the Central African Power Pool (CAPP) and the Eastern Africa Power Pool (EAPP) are dominated by hydropower and characterized by low electrification rates. In particular, EAPP is the most diverse one, as some of its members are entirely dependent on fossil fuels and others rely completely on renewable energies. In this context, an open modelling framework and input dataset have been developed for these three power pools, considering both the current (or near-future) situation and several long-term scenarios constrained by climate-related CO_2 limitations (Pavičević & Quoilin, 2020).

According to the results, in the reference scenario, capacity additions varying between 573 and 589 GW are anticipated by the year 2045, for an overall demand which is expected to grow by 16% by 2025, 89% by 2035, and 216% by 2045. Besides, a higher degree of interconnection could significantly reduce the load shedding and curtailment, especially in several countries with a very low generation capacity that do not share any cross-border lines with the neighbours (e.g. Central African Republic and South Sudan). The analyses also highlighted the dependence of the power sector on the availability of freshwater resources in the three power pools. Thus, differences between dry and wet years could vary the share of electricity coming from hydro units up to 5.2%, introduce changes in the operational costs around 1.4 billion € (or 3.28 €/MWh) and induce oscillations in CO_2 emissions up to 15 million tons per year. Besides, it is important to highlight that the water consumption of the power fleet in the NAPP is significantly low in relation to its water withdrawals (Figures CS3.1 and CS3.2). This is due to the large share of once-through cooling systems in NAPP, which increase the water temperature, but do not limit the quantity of water available for other users.

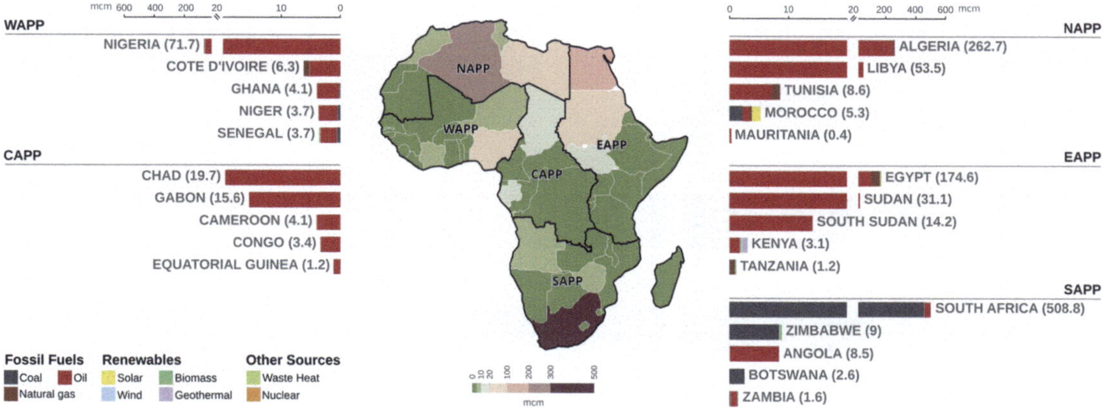

Figure CS3.1 Total water consumption (million cubic meters, mcm) per country by fuel type for the five power pools in 2016. The names and boundaries shown and the designations used on this map do not imply official endorsement or acceptance by the European Commission or the United Nations. The small islands are not considered in the study and the image reflects the study domain. *Source*: González Sánchez *et al.* (2020).

In this regard, a highly interconnected grid could reduce water withdrawals up to 50% across the three power pools (in comparison with the current system configuration) and water consumption between 50% in NAPP and 2% in EAPP. In addition, interconnections could reduce the price of electricity (between 2.7% in extremely wet seasons and 3.9% in extremely dry ones), as well as a higher integration of renewable sources. Furthermore, carbon emissions could be reduced by more than 32% (in comparison with the reference scenario).

CS3.6 CURRENT AND FUTURE PERSPECTIVES

Africa has the highest untapped hydropower potential in the world, as it is estimated that only 11% (37 GW) is currently used. Although in the last decade the hydropower installed capacity has increased at an average

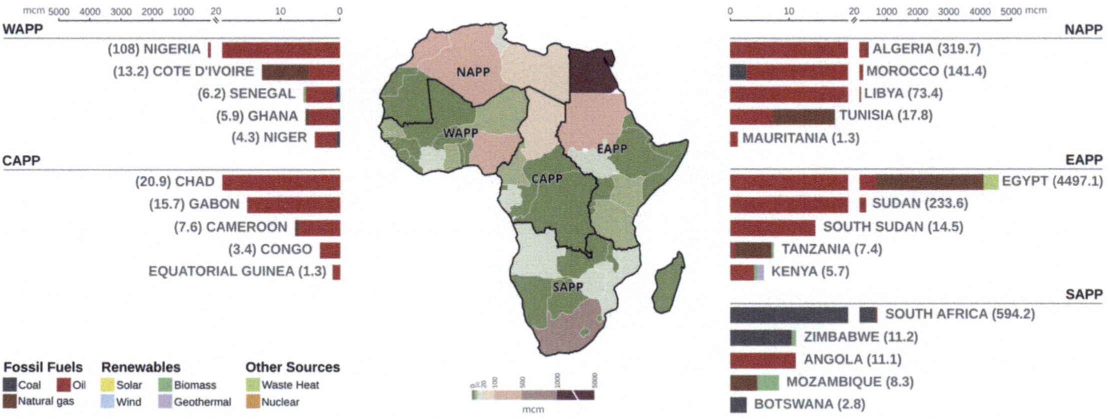

Figure CS3.2 Total water withdrawal per country by fuel type for the five power pools in 2016. The names and boundaries shown and the designations used on this map do not imply official endorsement or acceptance by the European Commission or the United Nations. The small islands are not considered in the study and the image reflects the study domain. *Source*: González Sánchez *et al.* (2020).

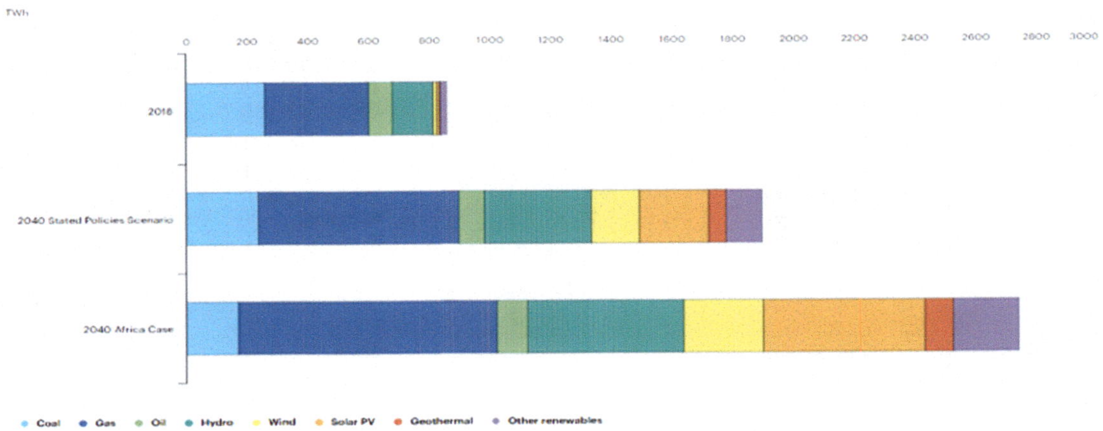

Figure CS3.3 Electricity generation in Africa by source (%) in 2018 and 2040 in the Africa case scenario. *Source*: IEA (2019), Africa Energy Outlook 2019, IEA, Paris, https://www.iea.org/reports/africa-energy-outlook-2019. All rights reserved; as modified by European Commission, Joint Research Centre.

annual rate of 4.4%, the effects of climate change and the ageing of the hydropower facilities (IHA, 2020) have prevented a similar growth in hydropower generation. In any case, hydropower plays an important role (Figure CS3.3) in all African regions except from North Africa, producing up to 17% of the total electricity in the continent, and expecting to grow in the following decades (generating more than 23% by 2040) (IEA, 2020).

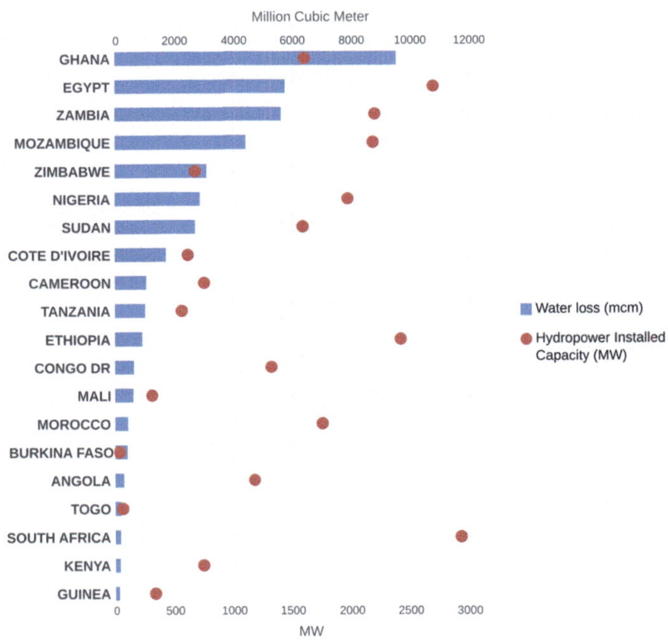

Figure CS3.4 Water loss from evaporation vs hydropower installed capacity for the top 20 countries with the highest water loss in 2016. *Source*: González Sánchez *et al.* (2020).

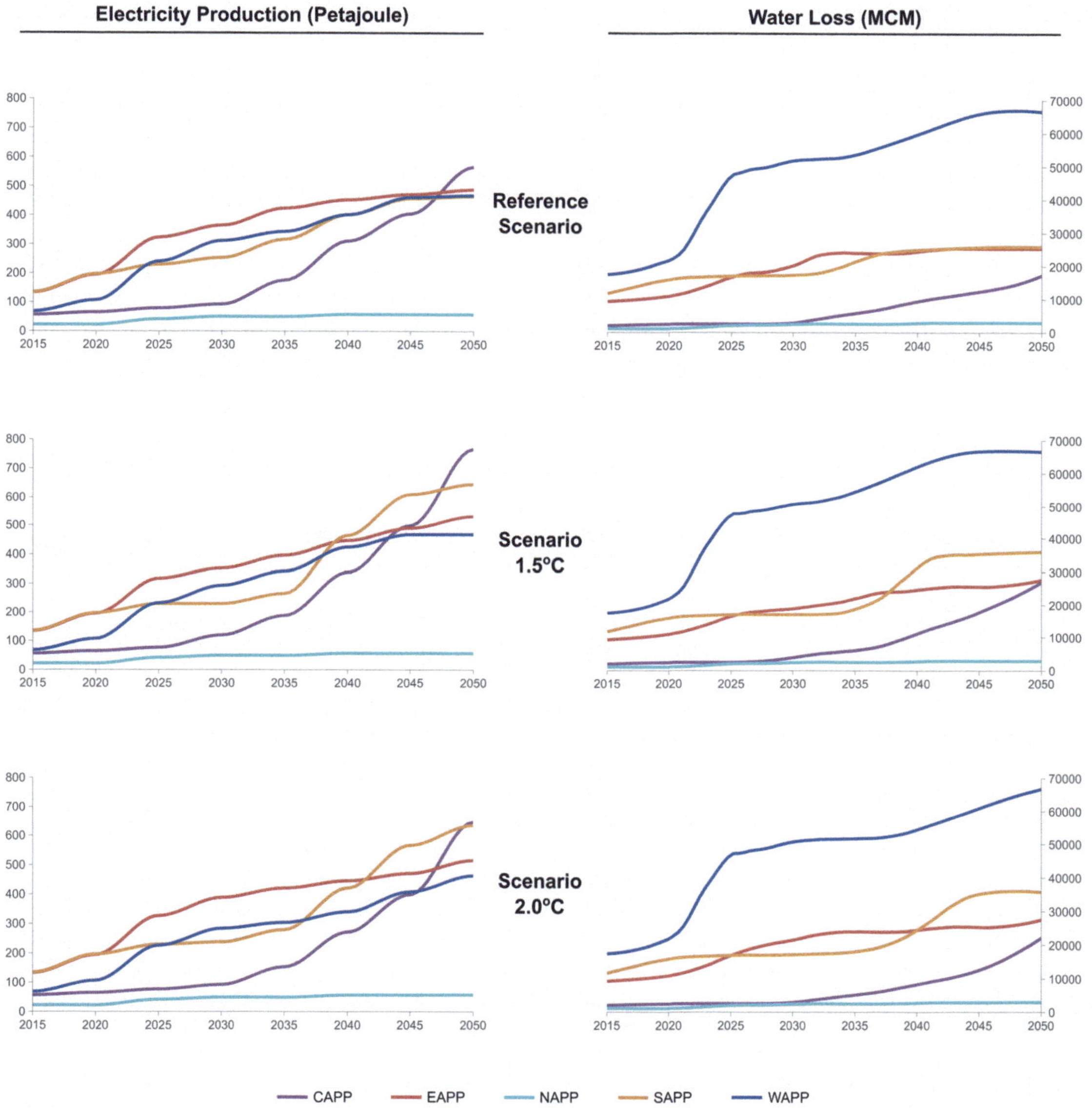

Figure CS3.5 Future projections of hydroelectricity production and water loss associated with hydropower. *Source*: González Sánchez *et al.* (2020).

However, the future energy mix will present significant differences at the regional level: while North, Eastern, and Southern Africa could obtain renewable energy from windpower, concentrating solar power (CSP) will be important specifically in North Africa, solar photovoltaics (PV) in both the Northern and Southern regions and geothermal sources in East Africa (IRENA, 2015). With regard to the latter, geothermal energy is expected to double its share by 2040 (from 2% of electricity generation in sub-Saharan countries (excluding South Africa) to 4%; IEA, 2019). Despite its marginal role at the continental level, this energy source could be highly relevant for some small African countries (e.g., Djibouti, Comores), as it might cover most of their electrical energy needs (Battistelli *et al.*, 2021).

To sum up, the use of non hydro renewable energies could offer an interesting alternative to fossil fuels, in order to contribute to the reduction of water use while fulfilling the increasing energy demand of the continent. As power systems pose impacts on both water resources' quantity and quality, the role of hydropower as a water user cannot be disregarded. In Africa, water losses linked to hydropower generation accounted for 42 billion cubic meters in 2016 (Figure CS3.4), whilst the correspondent to the combination of all the other fuel types was estimated at 1.2 billion cubic meters (Gonzalez Sanchez *et al.*, 2020). According to future projections evaporative losses in the African hydropower sector could undergo a significant increase in the next decades: they could range between 93.1 and 94.8 billion cubic meters by 2030 and between 139 and 160.7 billion cubic meters by 2050 (depending on the global warming scenario, Figure CS3.5; González Sánchez *et al.*, 2020).

REFERENCES

Battistelli A., Crestaz E. and Carmona-Moreno C. (2021). Status of geothermal industry in East African countries. ACEWATER2 report JRC121913.

Busch S., De Felice M. and Hidalgo Gonzalez I. (2020). Analysis of the water-power nexus in the Southern African power pool, EUR 30322 EN, Publications Office of the European Union, Luxembourg, ISBN 978-92-76-21015-3, doi: 10.2760/920794, JRC121329.

Conway D., Dalin C., Landman W. A. and Osborn T. J. (2017). Hydropower plans in eastern and Southern Africa increase risk of concurrent climate-related electricity supply disruption. *Nature Energy*, **2**, 946–953. https://doi.org/10.1038/s41560-017-0037-4

De Felice M., González Aparicio I., Huld T., Busch S. and Hidalgo González I. (2019). Analysis of the waterpower nexus in the west African power pool – water-energy-food-ecosystems project, EUR 29617 EN, Publications Office of the European Union, Luxembourg, ISBN 978-92-79-98138-8, doi: 10.2760/362802, JRC115157.

EC (2019). The European Green Deal. COM/2019/640 final.

EC (2020). Towards a comprehensive strategy with Africa. JOIN (2020) 4 final.

González Sánchez R., Hidalgo González I., Fahl F. and Seliger R. (2020). Current and projected freshwater needs of the African energy system, EUR 30278 EN, Publications Office of the European Union, Luxembourg, ISBN 978-92-76-19977-9, doi: 10.2760/808928, JRC120834.

Gonzalez Sanchez R., Seliger R., Fahl F., De Felice L., Ouarda T. B. M. J. and Farinosi F. (2020). Freshwater use of the energy sector in Africa. *Applied Energy*, **270**, 115171. doi: https://doi.org/10.1016/j.apenergy.2020.115171

Hidalgo Gonzalez I., De Felice M. and Busch S. (2021). Analysis of the water-power nexus in the African power pools. In: The African Networks of Centres of Excellence on Water Sciences Phase II (ACE WATER 2) WEFE Nexus assessment in Africa. JRC124127.

IEA (2019). Africa Energy Outlook 2019. World Energy Outlook Special report.

IEA (2020). Climate impacts on African hydropower.

IHA (2020). Hydropower Status Report. Sector trends and insights.

IRENA (2015). Africa 2030: roadmap for a renewable energy future, IRENA, Abu Dhabi.

Pavičević M. and Quoilin S. (2020). Analysis of the water-power nexus in the north, eastern and central African power pools, M. De Felice, S. Busch and I. Hidalgo Gonzalez, (eds), EUR 30310 EN, Publications Office of the European Union, Luxembourg, 2020, ISBN 978-92-76-20874-7, doi: 10.2760/12651, JRC121098.

World Bank (2013). World Bank's Thirsty Energy Initiative: http://www.worldbank.org/en/topic/water/ brief/water-energy-nexus.

Blended financing, underpinned by inter-sectorial risk management, enhances the commercial viability of nexus projects: lessons from the Songwe River Basin Development Programme

Z. Mpakama, E. A. Yaari and A. Earle

SIWI – Stockholm International Water Institute, Stockholm, Sweden

ABSTRACT

This section demonstrates the efficacy of using blended finance, underpinned by an inter-sectorial risk assessment approach, to improve the financial viability, as well as the environmental and social impacts of water, energy, and food-related infrastructure and projects. To achieve financial viability, a water, energy, and food nexus project needs to address these risks in an integrated manner that does not in effect prioritize one sector over another. The financing mechanisms available to infrastructure projects respond to varying degrees of risk inherent in the project. Blended financing makes use of commercial loans, concessionary loans and grants to cater for the various elements in the projects that carry differing risk levels. An integrated and robust risk management approach allows the blended finance structure to cater holistically to the financing needs of the project without excluding the less commercially viable components in favour of the cash-cows. Nexus projects with well-structured risk mitigation and financing will ultimately yield improved financial, environmental, and social returns for the project and community. The Songwe River Basin Development Programme (SRBDP) demonstrates how a nexus approach can address risks and increase a project's viability to attract financing.

Keywords: Nexus projects, blended financing, risk management

CS4.1 INTRODUCTION

Assessing the financial viability of water, energy, and food-related infrastructure and projects, is strongly correlated with risk management and mitigation. A project with a reduced risk profile has a higher chance of attracting financing from both public and private sources. Water, food, and energy-related infrastructure each present unique risk factors that need to be carefully understood and managed. To achieve financial viability, a water, energy, and food nexus project should be assessed taking into consideration the risks and opportunities of each component in an integrated manner. An integrated and robust risk management approach will ultimately yield improved financial, environmental, and social returns for the project and community. A look at the Songwe River Basin Development Programme

(SRBDP) under development in Tanzania and Malawi demonstrates how a nexus approach can address risks and increase a project's viability to attract financing using the blended financing approach.

CS4.2 BACKGROUND

Challenges surrounding financial viability are the single most identified reason for the slow pace of infrastructure development in developing economies and a bottleneck to attracting the private capital that is desperately needed to close the global infrastructure gaps and provide services needed by millions of people (Runde *et al.*, 2016). This paper demonstrates the importance of an integrated inter-sectorial risk assessment to improve the financial viability, as well as environmental and social impacts of water, energy, and food-related infrastructure and projects such as in the approach adopted by the SRBDP.

It is widely understood that the availability of investment resources is not the primary constraint to infrastructure development in developing economies, but rather that investment capital lacks quality projects that are investment ready (Runde *et al.*, 2016). It is project developers and sponsors who are responsible to their constituents to ensure that projects are developed to a financially viable and sustainable state, in contrast commercial debt providers are focused on the risk assessment inherent to the project to engage.

Early development stages of major water, energy, food infrastructure, and projects are focused on structuring the project to be financially viable, part of which involves the risk management and risk-sharing protocol of the project. The optimum risk-sharing protocol is characterized by assigning project risks to the party best suited to manage the identified risks. Projects with inadequate risk management and risk-sharing measures face the challenge of being unable to attract private capital. An integrated risk model that underpins the development of a sustainable project will take into account: the ownership structure; the communities where the project is located; the funding requirements and the project's ability to service the debt; technology, capacity of the infrastructure; environmental analysis; market analysis and contractual and institutional arrangements. The case of the SRBDP provides important insight into how risk management can appropriately be integrated into complex water, food, energy nexus projects.

CS4.3 THE SONGWE RIVER BASIN DEVELOPMENT PROGRAMME CASE STUDY

Challenges surrounding financial viability are the single most identified reason for the slow pace of infrastructure development in developing economies and a bottleneck to attracting the private capital that is desperately needed to close the global infrastructure gaps and provide services needed by millions of people (Runde *et al.*, 2016). This paper demonstrates the importance of an integrated inter-sectorial risk assessment to improve the financial viability, as well as environmental and social impacts of water, energy, and food-related infrastructure and projects such as in the approach adopted by the SRBDP.

It is widely understood that the availability of investment resources is not the primary constraint to infrastructure development in developing economies, but rather that investment capital lacks quality projects that are investment ready (Runde *et al.*, 2016). It is project developers and sponsors who are responsible to their constituents to ensure that projects are developed to a financially viable and sustainable state, in contrast commercial debt providers are focused on the risk assessment inherent to the project to engage.

Early development stages of major water, energy, food infrastructure, and projects are focused on structuring the project to be financially viable, part of which involves the risk management and risk-sharing protocol of the project. The optimum risk-sharing protocol is characterized by assigning project risks to the party best suited to manage the identified risks. Projects with inadequate risk management and risk-sharing measures face the challenge of being unable to attract private capital. An integrated risk model that underpins the development of a sustainable project will take into account: the ownership structure; the communities where the project is located; the funding requirements and the project's ability to service the debt; technology, capacity of the infrastructure; environmental analysis; market analysis, and contractual and institutional arrangements. The case of the SRBDP provides important insight into how risk management can appropriately be integrated into complex water, food, energy nexus projects. The Songwe River Basin (SRB) lies in the southwest of Tanzania and the north of Malawi. The SRB is, for the most part, a steep, mountainous catchment with an area of 4,243 square kilometres. About 55% of the catchment area is in Tanzania and 45% of the area is in Malawi. The Songwe River forms part of the border between Malawi and Tanzania over a distance of some 200 km.

CS4.4 SDGs FOUNDATIONS OF THE SRBDP

For nexus infrastructure development to be sustainable, financial structuring must be managed within the constraints of their finite nature and the interdependence of the ecosystem (Weitz *et al.*, 2014). Similarly to nexus projects, the SDGs are a set of integrated development goals intentionally designed to support and enhance one another. The SDGs aim to attain goals in one sector with positive or as little impact as possible on other sectors while relying on resource inputs from existing supplies without degrading the resources base and the underlying ecosystems.

Policy development within the water, energy, and food sectors deal with managing rapidly growing demand, finite supplies, and improving resource access for all. Without accounting for nexus-related risks and benefits, policies and programmes are less likely to be sustainable (Wakeford *et al.*, 2015). The guiding principle for the nexus approach and the SDGs is providing access to resources in an equitable and efficient manner and to ensure sustainability. The development of a hydro scheme such as that of the SRBDP adopts the nexus approach that links electricity generation through hydropower with irrigation schemes that support commercial and small-scale agricultural processes. New hydro schemes being developed and designed are adopting a more inclusive and sustainable approach to curb negative impact on the ecosystem and improve access for vulnerable and marginalized communities.

Policies that manage water, food, and energy and aim to operationalize the SDGs of Zero Hunger (SDG 2), Clean Water and Sanitation (SDG 6) and Affordable and Clean Energy (SDG 7) cannot be effective and impactful when developed and implemented in isolation. Competition for scarce resources poses a risk to ecosystems and resource security caused by degradation of the ecosystems and irreversible climate change effects. These sectors are inter-dependant, the water sector needs electricity for pumping, likewise, the energy sector needs water for electricity generation, and the food sector is both a consumer and source to energy and water sectors. A holistic policy framework approach can manage these complex relationships nestled amongst rapid population growth, changing economic conditions, and climate change (Weitz *et al.*, 2014).

In Figure CS4.1, we see that the SDG targets that underpin the water, energy, and food aim to ensure access to resources, improve efficient management and protection of resources, and provide long-term sustainability.

The SRBDP is one such project that aims at operationalizing the implementation of the SDGs. The project is implemented in a holistic manner with six sub-projects that will address water security, energy security, and food security of the Songwe River Basin and surrounding communities. The detailed description of the project components shows the SRBDP addresses food, water, and energy security and implementing a nexus project in a tangible manner.

- Access to affordable, reliable, and modern energy services
- Energy efficiency
- The share of renewable energy in the global energy mix

Energy Security

- Access to safe, nutritious and sufficient food
- Ending malnutrition
- Agricultural productivity
- Sustainable food production systems and resilient agricultural practices
- Genetic diversity of seeds, cultivated plants, farmed and domesticated animals and their related wild species, and access to fair and equitable sharing of the benefits from genetic resources

Food Security

- Universal and equitable access to safe and affordable drinking-water
- Water quality, wastewater, recycling, and reuse
- Water-use efficiency, sustainable withdrawals and supplies of freshwater, and reduction of water scarcity
- Integrated water resources management
- Access to sanitation and hygiene
- Protection and restoration of water-related ecosystems

Water Security

Source: *(Weitz, Nilsson & Davis, 2014, 41).*

Figure CS4.1 SDG goals that address energy, food, and water security.

(1) Lower Songwe Dam and Hydropower Plant (HPP) Project

The Lower Songwe Dam and Hydropower Plant Project is a multi-purpose dam, located in Ileje District, Tanzania and Chitipa and Karonga Districts, Malawi. The Project, with a capacity of 175 MW, is designed to fulfil the following purposes:

- Generate hydroelectric power by utilising the head between the water level impounded by the dam and the tailwater level in the river further downstream, at the location of the tailrace outlet;
- Protect the downstream reach of the Songwe River from floods particularly for the Lower Songwe Sub-Basin ('floodplain'), by retaining flood waters in the reservoir, and;
- Provide irrigation water to the floodplain.

(2) Lower Songwe River Tanzania Irrigation and Drainage Scheme

The Lower Songwe River Tanzania (LSRT) Irrigation and Drainage scheme is located in the lower part of the SRB on the left bank of the Songwe River and situated between the Songwe

River and the Kiwira River with its upper boundary near the town of Kasumulu just downstream of the Kasumulu Bridge. LSRT, as well as its related project in Tanzania, requires the construction of the Lower Songwe Dam for its water supply. The net irrigation area of LSRT is 3170 ha.

(3) Lower Songwe River Malawi Irrigation and Drainage Scheme

The Lower Songwe River Malawi (LSRM) Irrigation and Drainage scheme is located in the lower part of the SRB on the right bank of the Songwe River and situated between the Songwe River and the main highway from the town of Songwe towards Karonga. LSRM, as well as its related project in Malawi, requires the construction of the Lower Songwe Dam for its water supply. The net irrigation area of LSRM is 3019 ha.

The area planned for irrigation in both Tanzania and Malawi is currently a rain fed cropping area with the majority of the land under rice cultivation. The rice in this region is a traditional variety which is popular and fetches a good price. This would continue to be the main crop in the wet season, with irrigation being supplemental to the rainfall, and a good portion of it (60%) in the dry season, which will need constant irrigation. The other 40% will be typical upland crops of maize, groundnuts, cassava, and market vegetables.

(4) Stabilization of the Lower Songwe River

The project will also stabilize the Lower Songwe, where the meandering and instability of the Lower 39 km of the Songwe River is a natural phenomenon exacerbated by the deforestation of the river banks for agriculture and floods also influencing river instability. The river stabilization, prioritizing non-structural bank stability works, is intended to minimize the environmental damage to this important river as well as the costs, in terms of both initial investment and recurrent costs. Through stakeholder consultation, a collaborative approach to the design of the non-structural afforestation component was completed resulting in a set of village-based programmes for planting, maintenance, and management of the river stabilization.

(5) Water supply for Kasumulu and Songwe Towns and Lower Songwe Communities

The project will supply water to the towns of Kasumulu and Songwe and surrounding communities. The scheme is made up of:

- a raw water intake on the feeder canal for the LSRT I&D scheme 13.8 km downstream of the Lower Songwe Dam counter weir;
- a water treatment plant with a design capacity of 10,000 m^3/day with two parallel streams: of capacity 5000 m^3/day;
- a 3000 m^3 storage tank to work in combination with the existing tank;
- a distribution network to the Lower Songwe communities totalling 104 km of various pipe sizes.

(6) Water-related social infrastructure.

The water-related social infrastructure under the DDIP-SRBDP consists of four elements:

- Fisheries development;
- Water supply;
- Tourism development;
- Rural electrification.

CS4.5 DISCUSSION AND KEY FINDINGS

The SRBDP is made up of a diverse set of projects focusing on water resources, power generation, and social infrastructure projects. The projects vary in terms of risk profile and commercial potential. The socially oriented components have limited commercial potential, if any. The projects with revenue-generating capacity are the lower dam and associated hydropower plant; and the development of two irrigation schemes on the Songwe River.

To raise external commercial funding, the Lower Songwe Dam and Hydropower Plant (HPP) Project is considered as a standalone project due to the fact that it is able to generate distinct cash flows which can be readily ring-fenced and suitable for a typical project finance structure; that is, the financing of non-recourse/limited recourse long-term infrastructure where the debt and equity are repaid from the cashflows generated from the services provided by the infrastructure. Central to the project finance structure is the risk mitigation regime that is employed to transfer risk to the parties that can best manage them. The construction risk is addressed through an engineering, procurement and construction (EPC) contract where the contractor undertakes the performance, delay and cost overrun risks by constructing the facility on time, on budget and to specifications. Other risks, such as operating risk are addressed through the Operations and Maintenance (O&M) contract, cashflow risk for the HPP is managed by having a Power Purchase Agreement (PPA) in place that will undertake to buy the power generated at a stated price, eliminating the uncertainty presented by market risk.

As the SRBDP includes not only the development and implementation of the HPP, but an array of projects that aim to develop the basin and address food, water and energy security, the less financially viable projects also need to raise funds that can meet the ensuing risk levels. The nexus approach to the development of the SRBDP lends it to an array of funding options available and is representative of different levels of investor risk preference and appetite for the projects proposed. Two broad funding sources are suitable to implement this project, namely developmental capital, referring to finance provided at concessionary rates and a strong emphasis on the economic and social returns of the project; and commercial capital, where the focus is on earning market-related financial returns.

The SRBDP, as a whole, can be financially viable as a viable public–private partnership (PPP) project, if implemented under a blended finance approach that would involve significant concessional debt from Development Finance Institutions and/or Multilateral Development Banks, equity from the two Governments of Malawi and Tanzania, as well as commercial debt and equity from private investors.

The HPP project is financially viable with consistent returns, sufficient net profit, and able to service its debt commitments. The critical factor is structuring the right mix of the various forms of possible financing to ensure that the project generates the right returns to attract private investors.

For the SRBDP's irrigation projects, a mix of grant funding and concessional debt from development finance institutions would meet the capital expenditure needs and a nominal levy to fund the operational and recapitalization costs of the scheme. The positive economic and social viability of the irrigation schemes enhances the possibility of securing concessional debt finance for the capital cost of the project. The social projects can be funded through grants from donors as these are not commercially based initiatives.

CS4.6 CONCLUSIONS

Nexus projects, unlike traditional single sector infrastructure projects, are concerned with the sustainability of not one, but three different major sectors that often have conflicting aspects in implementation. This complicates matters for project sponsors as they must at all times consider the impact of each action on the holistic project and related sectors. To attract financing for nexus projects, robust, integrated risk mitigation systems must be put in place that carefully and clearly address all risk elements presented in the project, to ensure long-term sustainability.

A project's financial viability is based on strengthening a variety of factors, all put in place to mitigate the risks that emanate from implementing an infrastructure project. These factors, include: the capacity and financial strength of the project counterparts, for example, the sponsors, the contractors, the market or off-taker/s; the commercial strength of the business case and the ability for the project to service any debt and returns; and the socio-economic and environmental impacts of the project. A nexus project will deal with a multitude of stakeholders and parties in implementing the project; the effective management of interface risk is also crucial so that implementation is not delayed due to disputes. Clearly defined and communicated roles and responsibilities and associated risks will address this risk. Moreover, a very strong manager will have to be appointed who will manage each party during the implementation.

Nexus projects also address critical areas of socio-economic development, such as access to stable, reliable electricity, access to safe drinking water, building resistance to climate change, agricultural development, and by extension, contributing to poverty alleviation, job creation, improvement of livelihoods and development of small enterprises. Maximizing development impacts can improve sustainable financing of nexus projects through the ability to attract financing from varied sources – investment finance from commercial funders and private investors, to grants from donors, as seen in the case with the SRBDP.

REFERENCES

Runde D., Moser F. and Nealer H. E. (2016). Barriers to bankable infrastructure: incentivizing private investment to fill the global infrastructure gap, Centre For Strategic & International Studies.

Wakeford J., Kelly C. and Lagrange S. M. (2014). Mitigating Risks & Vulnerabilities in the Energy-Food-Water Nexus in Developing Countries. Sustainability Institute and School of Public Leadership, Stellenbosch University.

Weitz N., Nilsson M. and Davis M. (2014). A nexus approach post the 2015 agenda: formulating integrated water, energy and food SGDs. *SAIS Review of International Affairs*, **34**, 37–50.

WEFE Nexus regional studies

Water energy and food ecosystem vulnerability in the Senegal River

M. Pastori, P. Marcos-Garcia, G. Umlauf, J. Koundouno, L. Cattaneo, E. Cordano and C. Carmona

European Commission, Joint Research Centre, Directorate D – Sustainable Resources, D2 – Water and Marine Resources Unit, Ispra (Va), Italy

ABSTRACT

African developing countries are confronted with the challenge of effectively managing natural resources to achieve higher outcomes in several sectors, such as pointed out in the SDGs, while ensuring sustainability and environmental protective solutions. This challenge requires considerable efforts in transboundary river basins, such as the Senegal River Basin (SRB), due to complex and cross-sectoral technical and political realities. To this scope in this report a preliminary identification of WEFE Nexus-related issues in the SRB are identified and summarized. These issues also include interactions between key sectors (i.e., water, energy, agriculture, and environment), namely:

- hydropower development using multi-purpose infrastructure and the social and environmental impacts associated;
- high climate variability and its impacts on the production system, nature and people, particularly on the poorest and rain dependent socio-economic communities;
- improvement of production systems (irrigation, rain-fed, flood recession agriculture) to increase crop production and food security;
- navigation improvement to enhance commerce and support development; environmental protection and safeguarding of specific ecosystems (such as the Delta);
- water quality and impacts on water-related diseases and socio-economic dynamics.

Keywords: WEFE Nexus, climate, hydropower, agriculture, Senegal River Basin

CS5.1 INTRODUCTION

African developing countries are confronted with the challenge of effectively managing natural resources to achieve higher outcomes in several sectors, such as pointed out in the SDGs (United Nations (UN), 2015), while ensuring sustainability and environmental protective solutions.

This challenge requires even more effort in transboundary river basins, where solutions should be balanced not only across competing sectors and scales but also taking into account the specific and eventually different and competing development objectives of the riparian countries. In this regard, a sound, integrated and inclusive transboundary management approach is clearly needed to effectively address coming and future challenges at river basin scale, while ensuring at the same time sustainability.

In recent years, the WEFE Nexus approach has taken a centre stage as an integrative approach meant to integrate both management and governance across the multiple sectors that its four components involve (e.g. agriculture, food, fishing, livestock, energy, forest protection, water quality, etc.) and its concept has rapidly expanded (Albrecht *et al.*, 2018; Keairns *et al.*, 2016). The use of such approach should overcome the traditional view of single sectors as separate entities, featuring them instead as complex and inextricably systems (EC, 2019) where interlinkages and feedbacks across sectors should be carefully addressed. Currently, the European Union (EU) is actively cooperating with the African Union (AU) in several policy initiatives framing the demand for the WEFE Nexus approach to water development in Africa. In this context, the WEFE Senegal project (APPUI A LA GESTION DES RESSOURCES EN EAU ET DU NEXUS EAU-ENERGIE-AGRICULTURE DANS LE BASSIN DU FLEUVE SENEGAL), which is funded by the European Union is being implemented by the the Joint Research Centre (JRC) of the European Commission and the AICS[1] in collaboration with the Organisation pour la Mise en Valeur du fleuve Sénégal (OMVS),[2] the Directorate-General of International Partnership (DG INTPA) and the European Union Delegation in Dakar (Senegal).

The main goal of this project is to contribute to and strengthen the WEFE framework in the Senegal River Basin (SRB), in order to improve the understanding of the interactions between water resources management, climate change, and the evolution of agricultural activities in a rural economy. Concretely, the scientific component of the project aims to:

(a) Strengthen technical/scientific knowledge on relevant components and issues identified in the SRB, in collaboration with local/regional technical actors;
(b) Promote sustainable management measures in coherence with the policies and governance of the basin, taking into account regional (Water Management Master Plan, Common Energy Policy, Energy Transport Master Plan, Regional Action Plan for the Improvement of Irrigated Crops, Strategic Environmental Action Plan, etc.) and national policies;
(c) Provide support for the assessment and evaluation of alternative measures and solutions as proposed by the OMVS.

CS5.2 THE SENEGAL RIVER BASIN

The Senegal River is the second longest river (1800 km) in West Africa and its transboundary drainage basin covers about 410,000 km^2, over Guinea, Mali, Mauritania, and Senegal (10, 54, 26, and 15%, respectively). Born in the Fouta Djallon massif in Guinea, the Senegal River travels across Guinea and Mali and, after the confluence of the Bafing, Bakoye, and Falémé rivers, traces the border between Mauritania and Senegal until it meets the Atlantic ocean, near Saint-Louis in Senegal (Table CS5.1). The journey of the river constitutes a lifeline for 7.5 million people of the basin (16% of the riparian countries' population) but also for the economy of the riparian countries and the region. Due to the high

[1]AICS: Italian Agency for development cooperation. Web: https://dakar.aics.gov.it/.

[2]OMVS :Organisation pour la mise en valeur du fleuve Sénégal. Web: http://www.omvs.org/.

Table CS5.1 Main SRB characteristics.

SRB Factsheet		
Surface area		340,000 km^2
Precipitation	Mean	550 mm/yr
	Range	200–1500 mm/yr
Discharge (OMVS data)		
Bafing Makana	<80s	330 m^3/s
	>2010	267 m^3/s
Bakel	<80s	778 m^3/s
	>2010	611 m^3/s
Population (OMVS-SDAGE)	Habitants (2015)	7,980,000
	Density	23–25 hab/km^2
	Rural	29.6%
	Growth	2.8%
	Habitants (2025)	9,000,000
Agriculture		
Harvested area	OMVS-SDAGE	21,500 km^2
Cropland	ESA 2014	79,000 km^2
Dominant crops	OMVS-SDAGE	Millet, maize, sorghum (51%)
		Rice (7%), pulses (12%), oils
Irrigated area	OMVS-SDAGE	75,600 ha
Livestock	(from Gilbert *et al.*, 2014)	
	Cattle	16 million
	Goats-sheeps	46.5 million
	Poultry	96.6 million

dependency of the main livelihoods in SRB on water (agriculture, livestock, fisheries), around 85% of its population lives close to the river (UN, 2003).

In 1972, Senegal, Mauritania, and Mali decided to join their efforts in developing the basin through the establishment of the OMVS, which is considered as an example of transboundary cooperation due to the effective implementation of the principle of equitable sharing among member states (regarding both the ownership of hydraulic infrastructure and the benefits associated with water resources at the national level). In 2006, Guinea joined the OMVS. Despite the efforts made by the four riparian countries during the last three decades with regard to the basic dimensions of human development, they still show values of the Human Development Index (HDI) among the lowest of the world (ranking in the interval position 159–182 for a total of 189 countries) and are catalogued as least developed countries (UNDP, 2018; UN, 2021). Therefore, the development of the basin is of vital importance for the four countries, which face up significant challenges due to biophysical and socioeconomic determinants, such as the climate variability increase in the region, a high population growth rate (around 3% per year) or the lack of job opportunities (Figure CS5.1).

The SRB is highly vulnerable to climate variability and changes, due to the great interdependence between climate and socioeconomic activities, and it could be further challenged by the increasing pressures posed by its population dynamics on natural resources, the subsequent changes in land use and the competition among sectors and users. West Africa has suffered a severe drought which started in the late 1960s and continued for more than 40 years (e.g., Bodian, 2014; Nicholson *et al.*, 2000), impacting severely the discharge of the largest rivers in the region (Senegal and Niger rivers). However, in the case of the Senegal river, several authors have identified a turning point in both rainfall and river discharge between 1993 and 1994, when these variables started to show some signs of recovery (Bodian *et al.*, 2020; Hubert *et al.*, 2007). Concretely, there is evidence that the recovery of annual rainfall in the SRB is leading to the improvement of surface water availability (Bodian *et al.*, 2020), even if the persistence

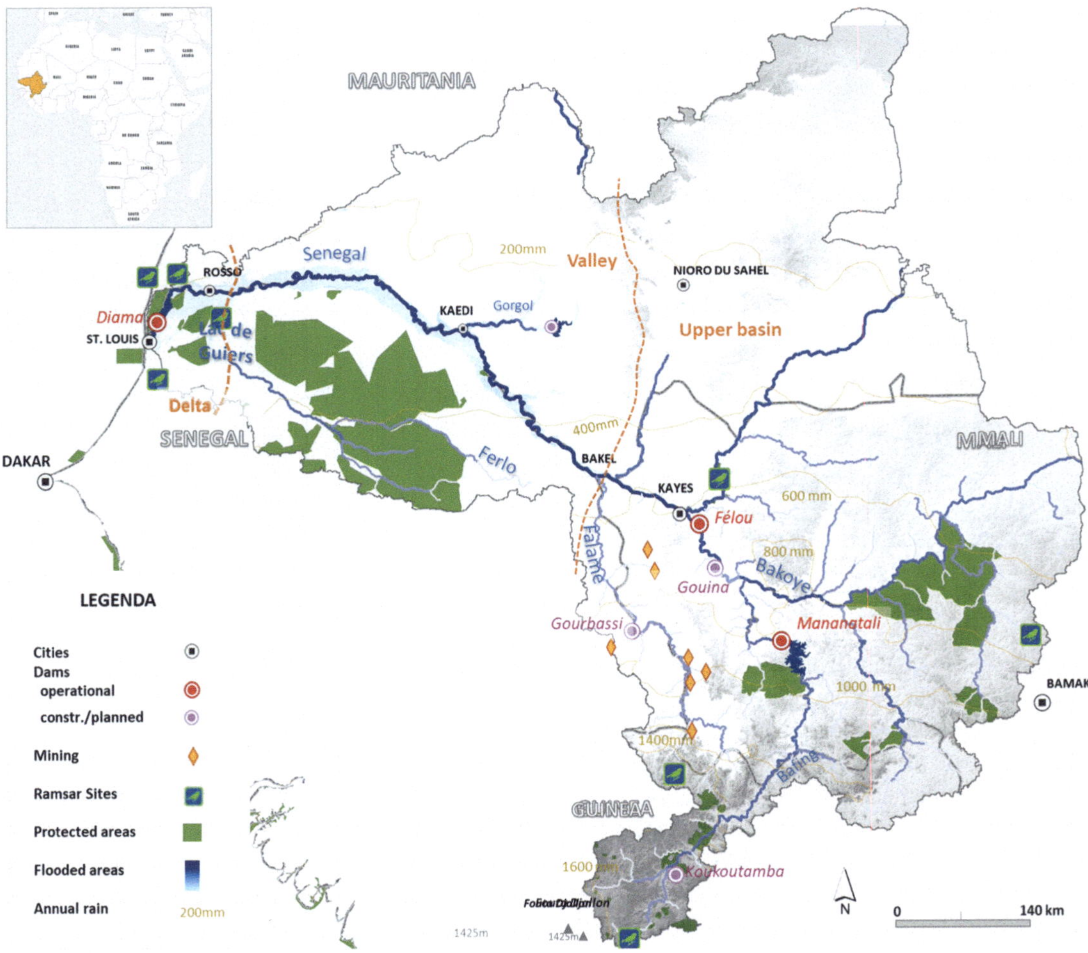

Figure CS5.1 The Senegal River Basin.

of this change should still be verified. Concerning trends in annual discharge river flow data, several studies (Bader, 2015; Pastori *et al.*, 2020) evidenced the presence of statistically significant positive trends between 1980s and 2020.

In any case, future water resource availability is not easy to predict, as regional climate models (RCMs) outputs often do not converge in Western Africa and the complexity of hydrosystems' responses in this semiarid region cannot be disregarded (Karambiri *et al.*, 2011). Concretely, the so-called Sahelian Paradox portraits the increase of runoff which took place in Sahelian basins for three decades in spite of the persistent drought, and it has been largely attributed to dramatic changes in land use (e.g., Descroix *et al.*, 2009). Climate variability is also specifically important due to its influence on agricultural production (above all when it comes to rainfed systems, which are the most widespread accounting for more than 90% of cropland and an important source of food security self-sufficiency for a great part of the rural population) and is a supporting analysis helping stakeholders in taking appropriate measures to reduce risks and impacts. Climate variability analysis developed in the framework of the WEFE Senegal project (Cattaneo *et al.*, 2019) focused on the assessment of several indicators, such as precipitation deficit, heat waves magnitude index, the Standardized Precipitation Index (SPI) and the characterization of dry spells. The results showed higher precipitation deficits and variability specifically in areas with lower precipitation, that are strongly dependent on rainfall for both human and livestock water supply, as well as for sustaining rainfed agriculture and pastureland (see e.g., Figure CS5.2: monthly deficit and average precipitation for only 2 months during rainy season are here).

There is a high hydropower potential in the basin and even if currently only two plants are being exploited (one under development), the four riparian countries and the OMVS have planned to increase the number of reservoirs, in order to meet the expected growing demands as well as to regulate the high inter- and intra-annual water availability of the basin (Tractebel, OMVS, 2013). The existing dams play the main role in river flow regulation, hydroelectricity production and development and control of irrigation (including flood recession control). In the middle valley and delta, agriculture, pastoralism, and fishing are the main activities and employ a large part of the working population, therefore providing most of the household income (even if clear and detailed disaggregated data are not available so far). All this region is poor and extremely dependent on the flood-related cropping activities in the depressions along the river for food security (Diouf, 2015).

If properly designed, the development of hydraulic infrastructures could act as a buffer against climate variability. Raso *et al.* (2019) performed an ex-ante economic evaluation of the Manantali and Diama dams and highlighted their ability to partially hedge natural variability and, hence, their economic potential under changing conditions (for both operational and structural changes). Initially, the OMVS ambitions in relation to the implementation of these schemes were high: improving food security through the development of irrigated agriculture (240,000 ha in Senegal, 126,000 ha in Mauritania and 9000 in Mali), supplying the three fast growing capitals with electricity (considering an economic viability level of power production of 800 GwH/yr) and enhancing the river's navigability. However, years later the results are mixed: irrigated agriculture has developed at a slower pace than expected with mixed and controversial socio-economic and environmental results (Manikowski & Strapasson, 2016) and navigability has not undergone a real increase as far as Mali, although the goals for starting the hydroelectric power production were fulfilled by the end of 2002 (Mietton *et al.*, 2007). Currently, the OMVS still hopes to create a continuous and lasting navigable waterway of more than 900 km between Saint-Louis (Senegal) and Ambidedi (Mali) (IFGR, 2018). In any case, future developments in the basin should consider the existence of different perspectives among the riparian countries and the specific challenges of the different regions of the basin. Tilmant *et al.* (2020) performed a probabilistic trade-off analysis between competing uses and evidenced the existence of two main coalitions, contending and competing for

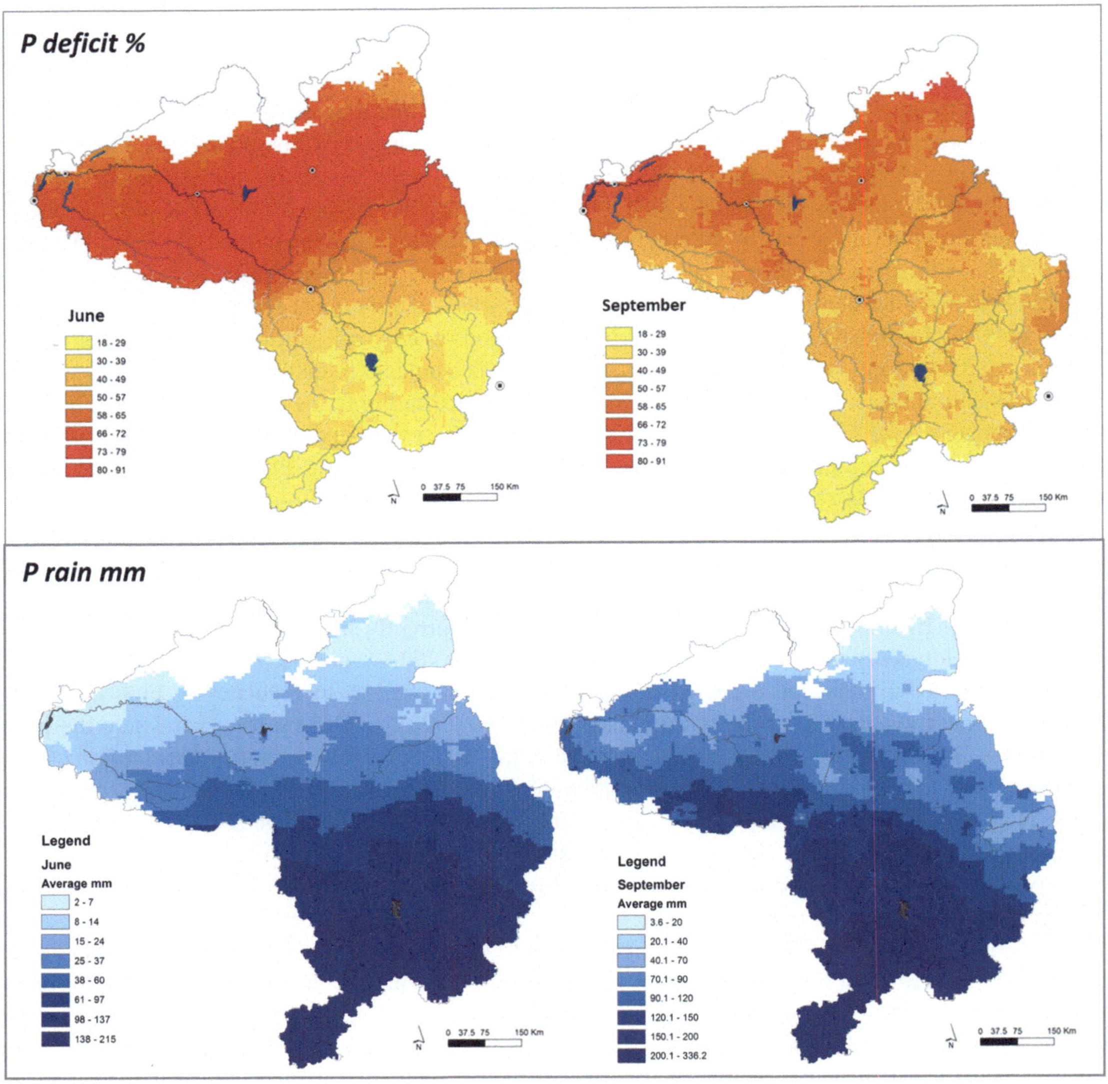

Figure CS5.2 Precipitation deficit (% divergence with respect to mean climatology) with a return period of 20 years (top) and average monthly precipitation (bottom) over the SRB for June and September.

specific advantages and developments strategies. While upstream countries (mainly Mali and Guinea) are highly interested in hydropower services, downstream countries (mainly Senegal and Mauritania) prioritize food production and ecosystem services in the valley and Delta areas (Tilmant *et al.*, 2020). Also the strategic action plan 2017–2037 of OMVS and its roadmap present key actions to be implemented for the management of priority environmental issues in the three zones (Fouta Djallon, Upper basin in Mali, Delta – targeted by OMVS and corresponding to the water planning and management plans elaborated in 2013) and for operationalizing the Water Charter 2002, the legal framework of reference for the Nexus, along with specific actions on gender.

Besides, a limited understanding of the interdependences among the components of the WEFE Nexus in the SRB has led to limited informed decision-making processes and multiple undesirable impacts following the implementation of Manantali and Diama dams. For example, the development of irrigated agriculture thanks to reservoirs might not be able to replace the total losses induced in flood-recession farming, which plays a significant role in food security in the area (Barbier & Thompson, 1998; Manikowski & Strapasson, 2016). According to Sall *et al.* (2020), current reservoir operation rules in the SRB assign the lowest priority to flood-recession agriculture, reducing the flooded area and the duration of the flood and, thereby, threatening the future of this farming practice and other types of livelihoods or ecosystem services (e.g. freshwater fish production, estuarine/marine fishery nursery grounds, and dry season forage). Other negative outcomes of water regulation and changes of the hydraulic regime in the SRB included deforestation, massive population displacements, groundwater and fishing depletion, increase of the number of invader aquatic species, and the rise of waterborne diseases (DeGeorges and Reilly, 2006; Mietton *et al.*, 2007; Diessner, 2012). With regard to this last issue, dam-induced changes in salinity and the provision of new suitable areas for the development of freshwater snails (irrigation channels, rice fields) have resulted in the spread of schistosomiasis throughout the Middle and Lower Valleys of the SRB, subsequently increasing its prevalence and intensity among the human population (Picquet *et al.*, 1996; Southgate, 1997). Besides, Dia *et al.* (2008) demonstrated the changes on the composition of malaria transmission, vectorial system and epidemiology due to the implementation of dams in the Senegal river.

Specifically, combating water-borne diseases is one of the key challenges in the SRB, due to the rapid increase in the prevalence of multiple diseases that were already present in the area (e.g. malaria, urinary schistosomiasis, diarrhoea, intestinal parasitic diseases), and the appearance and subsequent expansion of the previously cited intestinal schistosomiasis, a much more dangerous form of the disease (which particularly affects agricultural and fishing populations and impairs productivity, due to its debilitating nature) (Monde, 2016). Regarding schistosomiasis, since a while hybrid forms of the disease that jump in between man, livestock, rodents, etc. are observed, adding another level of complexity to the attempt of interrupting the transmission chains. In addition, the hybridization man/livestock imposes an economic dimension of Schistosomiasis on the farming sector (besides the economic consequences arising from the impacts on human productivity, education, etc.) (Léger *et al.*, 2020).

Besides, children in the SRB show the highest prevalence of another waterborne gastrointestinal parasite (Blastocystis) worldwide (El Safadi *et al.*, 2014) due to poor hygiene, sanitation, and water supply from unsafe sources, along with close contact with domestic animals and livestock. According to the individual country profiles, the environmental risk factors account for about one third of the total mortality in all cases, and specifically poor water quality issues clearly dominate the health impacts in all riparian countries, closely followed by risk factors linked to poor indoor air quality (WHO, 2009) (Table CS5.2).

Table CS5.2 Environmental burden of disease for selected risk factors.

WHO Statistics (2004)	Deaths/Year							
	Guinea		Mali		Mauritania		Senegal	
Risk factor	*n*	%	*n*	%	*n*	%	*n*	%
Water, sanitation, and hygiene	9600	60.4	22,600	58.1	2300	62.2	12,900	81.1
Indoor air	5700	35.8	15,300	39.3	1200	32.4	6300	39.6
Outdoor air	600	3.8	1000	2.6	200	5.4	1800	11.3

Even if less important and requiring a low priority in the basin if compared with vector and water-borne diseases and drinking water quality issues, also water pollution with chemicals could be a potential problem in certain zones. The growing agricultural development is increasing nitrate concentrations, although nutrient concentrations in the SRB reveal a relatively low human impact (in comparison to rivers in developed countries, where nutrient water pollution is still a major environmental issue for several rivers (Malagó *et al.*, 2019; Mbaye *et al.*, 2016). Besides, Troussellier *et al.* (2004) found eutrophication issues in the Senegal River estuary, due to the combination of numerous pollution sources in the area of Saint Louis city and the limited renewal of estuarine water. Regarding pollution due to heavy metals, El Mahmoud-Hamed *et al.* (2019) assessed the presence of cadmium, lead, and mercury in freshwater fish in the Senegal River in Mauritania and warned that they could pose a health risk in certain locations, such as in Kaédi and Boghé, due to high exposure (eating) frequency. Land-based plastic waste inputs into the ocean is another rising issue, as mismanaged plastic waste generally ends up in drains, landfills, and inland water bodies and finally into the marine area, although most Western African countries are implementing policies to curb it (Adam *et al.*, 2020). In this regard, it is estimated that only in Senegal (where the river mouth is located) more than 250,000 tonnes of plastic waste were mismanaged in 2010 (and thereby susceptible of becoming marine debris), and this amount was expected to triplicate in 2025 (Jambeck *et al.*, 2015).

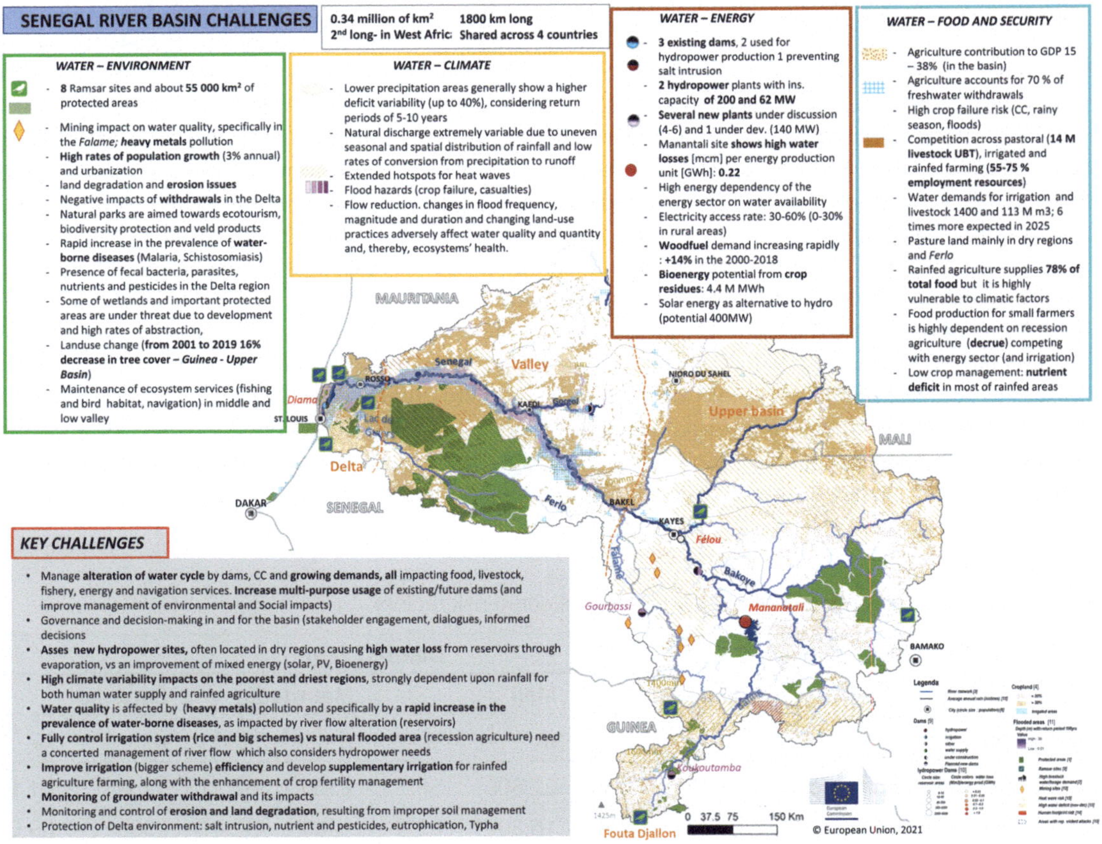

Figure CS5.3 SRB challenges.

CS5.3 KEY WEFE NEXUS CHALLENGES

Supplying sufficient water, food, and energy while maintaining environmental sustainability is a growing challenge due to rapid population growth, changing lifestyles, ecosystem degradation, increasing water scarcity, political rather than analysis-based, cross-sectoral inclusive decision making, and an uncertain future climate. By analysing OMVS documents, and scientific contributions most urgent challenges to be addressed in the basin are summarized here. These issues also include interactions between key sectors (i.e., water, energy, agriculture, and environment), concretely: hydropower development using multi-purpose infrastructure; high climate variability impact on the poorest and rain-dependent socio-economic communities; improvement of irrigation systems to increase crop production and food security; flood recession agriculture; navigation improvement to enhance commerce and development; environmental protection and safeguarding of specific ecosystems (such as the Delta); water quality and impacts on water-related diseases; and finally monitoring of groundwater withdrawals. Some key factors and challenges across the SRB have been summarized and highlighted in Figure CS5.3.

ACKNOWLEDGEMENTS

This work is a part of the 'WEFE Senegal project – Appui a la gestion des ressources en eau et du nexus eau-énergie-agriculture dans le bassin du fleuve Sénégal' funded by the European Commission.

REFERENCES

Adam I., Walker T. R., Bezerra J. C. and Clayton A. (2020). Policies to reduce single-use plastic marine pollution in West Africa. *Marine Policy*, **116**, 103928, ISSN 0308-597X.

Albrecht T. R., Crootof A. and Scott C. A. (2018). The water-energy-food nexus: a systematic review of methods for nexus assessment. *Environmental Research Letters*, **13**. doi: 10.1088/1748-9326/aaa9c6

Bader J. C. (2015). Monographie hydrologique du fleuve Sénégal, EAN13 (CD-ROM): 9782709918855. 30. IRD Editions, Montpellier.

Barbier E. and Thompson J. R. (1998). The value of water: floodplain versus large-scale irrigation benefits in northern Nigeria. *Ambio*, **27**(6), 434–440.

Bodian A. (2014). Caractérisation de la variabilité temporelle récente des précipitations annuelles au Sénégal (Afrique de l'Ouest), *Physio-Géo*, **8** | -1, 297–312.

Bodian A., Diop L., Panthou G., Dacosta H., Deme A., Dezetter A., Ndiaye P. M., Diouf I. and Vischel T. (2020). Recent trend in hydroclimatic conditions in the Senegal River Basin. *Water*, **12**, 436.

Cattaneo L., Pastori M., Cordano E., Crestaz E., Seliger R., Koundouno J., Bausa Lopez L. and Carmona C. (2019). Title Applications and results for the E-Nexus decision support system for the WEFE Senegal project, European Commission, Ispra, JRC119391.

DeGeorges A. and Reilly B. K. (2006). Dams and large scale irrigation on the Senegal River: impacts on man and the environment. *International Journal of Environmental Studies*, **63**(5), 633–644. doi: 10.1080/00207230600963296

Descroix L., Mahé G., Lebel T., Favreau G., Galle S., Gautier E., Olivry J. C., Albergel J., Amogu O., Cappelaere B., Dessouassi R., Diedhiou A., Le Breton E., Mamadou I. and Sighomnou D. (2009). Spatio-temporal variability of hydrological regimes around the boundaries between Sahelian and Sudanian areas of West Africa: A synthesis. *Journal of Hydrology*, **375**(1–2), 90–102. https://doi.org/10.1016/j.jhydrol.2008.12.012

Dia I., Konate L., Samb B., Sarr J. B., Diop A., Rogerie F., Faye M., Riveau G., Remoue F., Diallo M. and Fontenille D. (2008). Bionomics of malaria vectors and relationship with malaria transmission and epidemiology in three physiographic zones in the Senegal River Basin. *Acta Tropica*, **105**(2), 145–153. https://doi.org/10.1016/j.actatropica.2007.10.010.

Diessner C. (2012). Dam complications in Senegal: how river dams may hurt more than help vulnerable populations in water-stressed regions. *Journal of Environmental and Sustainability Law*, **19**, 247. Available at: https://scholarship.law.missouri.edu/jesl/vol19/iss1/10 (last accessed 19 Jul 2021).

Diouf Y. (2015). Mémoire de fin d'études. «Étude socioéconomique de la vulnérabilité/résilience des exploitations agricoles familiales de la vallée du fleuve Sénégal». Université de Thiès – École Nationale Supérieure d'Agriculture (ENSA).

EC (2019). Position Paper on Water, Energy, Food, and Ecosystem (WEFE) Nexus and Sustainable development Goals (SDGs). C. Carmona-Moreno, C. Dondeynaz, M. Biedler (eds), EUR 29509 EN, Publications Office of the European Union, Luxembourg, 2019, ISBN 978-92-79-98276-7. doi: 10.2760/5295, JRC114177.

El Mahmoud-Hamed M. S., Montesdeoca-Esponda S., Santana-Del Pino A., Zamel M. L., Brahim M., T'feil H., Santana-Rodiguez J. J., Sidoumou Z. and Sidi'Ahmed-Kankou M. (2019). Distribution and health risk assessment of cadmium, lead, and mercury in freshwater fish from the right bank of Senegal River in Mauritania. *Environmental Monitoring and Assessment*, **191**, 493. https://doi.org/10.1007/s10661-019-7627-5.

El Safadi D., Gaayeb L., Meloni D., Cian A., Poirier P., Wawrzyniak I., Delbac F., Dabboussi F., Delhaes L., Seck M., Hamze M., Riveau G. and Viscogliosi E. (2014). Children of Senegal River Basin show the highest prevalence of Blastocystis sp. ever observed worldwide. *BMC Infectious Diseases*, **14**, 164. https://doi.org/10.1186/1471-2334-14-164.

Gilbert M., Nicolas G., Cinardi G., Van Boeckel T.P., Vanwambeke S.O., Wint G.R.W. and Robinson T.P., 2018. Global distribution data for cattle, buffaloes, horses, sheep, goats, pigs, chickens and ducks in 2010. *Scientific Data*, **5**, 1–11. doi:10.1038/sdata.2018.227

Hubert P., Bader J.-C. and Bendjoudi H. (2007). One century of Senegal River annual discharges. *Hydrological Sciences Journal*, **52**(1), 68–73. doi: 10.1623/hysj.52.1.68.

IFGR (2018). Making the Senegal River navigable to transform it into an instrument for economic development and integration. Appeals and recommendations. Initiatives for the Future of Great Rivers (IFGR), 6th session, 9–13 April 2018.

Jambeck J. R., Geyer R., Wilcox C., Siegler T. R., Perryman M., Andrady A., Narayan R. and Law K. L. (2015). Plastic waste inputs from land into the ocean. *Science*, **347**(6223), 768–771. doi: 10.1126/science.1260352.

Karambiri H., García Galiano S. G., Giraldo J. D., Yacouba H., Ibrahim B., Barbier B. and Polcher J. (2011). Assessing the impact of climate variability and climate change on runoff in West Africa: the case of Senegal and Nakambe River basins. *Atmospheric Science Letters*, **12**, 109–115. https://doi.org/10.1002/asl.317.

Keairns D. L., Darton R. C. and Irabien A. (2016). The energy-water-food nexus. *Annual Review of Chemical and Biomolecular Engineering*, **7**, 239–262. doi: 10.1146/annurev-chembioeng-080615-033539.

Léger E., Borlase A., Fall C. B., Diouf N. D., Diop S. D., Yasenev L., Catalano S., Thiam C. T., Ndiaye A., Emery A., Morrell A., Rabone M., Ndao M., Faye B., Rollinson D., Rudge J. W., Sène M. and Webster J. P. (2020). Prevalence and distribution of schistosomiasis in human, livestock, and snail populations in northern Senegal: a One Health epidemiological study of a multi-host system. *The Lancet Planetary Health*, **4**, e330–e342. doi: 10.1016/S2542-5196(20)30129-7.

Malagó A., Bouraoui F., Grizzetti B. and De Roo A. (2019). Modelling nutrient fluxes into the Mediterranean Sea. *Journal of Hydrology: Regional Studies*, **22**, 100592. doi: 10.1016/j.ejrh.2019.01.004.

Manikowski S. and Strapasson A. (2016). Sustainability assessment of large irrigation dams in Senegal: a cost-benefit analysis for the Senegal River Valley. *Frontiers in Environmental Science*, **4**, 18. doi: 10.3389/fenvs.2016.00018.

Mbaye M. L., Gaye A. T., Spitzy A., Dähnke K., Afouda A. and Gaye B. (2016). Seasonal and spatial variation in suspended matter, organic carbon, nitrogen, and nutrient concentrations of the Senegal River in West Africa. *Limnologica*, **57**, 1–13. https://doi.org/10.1016/j.limno.2015.12.003.

Mietton M., Dumas D., Hamerlynck O., Kane A., Coly A., Duvail S., Pesneaud F. and O. Baba M. L. (2007). Water management in the Senegal River Delta: a continuing uncertainty. *Hydrology and Earth System Sciences Discussion*, **4**, 4297–4323.

Monde C. (2016). Impact of Natural and Anthropogenic Factors on the Trophic Interactions of Molluscivores and Schistosoma Host Snails. Wageningen University. https://edepot.wur.nl/378592.

Nicholson S. E., Some B. and Kone B. (2000). An analysis of recent rainfall conditions in West Africa, including the rainy seasons of the 1997 El Niño and the 1998 La Niña years. *Journal of Climate*, **13**(14), 2628–2640.

Pastori M., Cattaneo L., Koundouno J. and Carmona C. (2020). e-Nexus Decision Support System First development and application in the context of the WEFE Senegal Project – Hydrological component. European Commission, Ispra. WEFE Senegal Internal Report.

Picquet M., Ernould J. C., Vercruysse J., Southgate V. R., Mbaye A., Sambou B., Niang M. and Rollinson D. (1996). The epidemiology of human schistosomiasis in the Senegal River basin. *Transactions of the Royal Society of Tropical Medicine and Hygiene*, **90**(4), 340–346.

Raso L., Barbier B. and Bader J.-C. (2019). Modeling dynamics and adaptation at operational and structural scales for the ex-ante economic evaluation of large dams in an African context. *Water Resources and Economics*, **26**, 100125.

Sall M., Poussin J.-C., Bossa A. Y., Ndiaye R., Cissé M., Martin D., Bader J.-C., Sultan B. and Ogilvie A. (2020). Water constraints and flood-recession agriculture in the Senegal River valley. *Atmosphere*, **11**, 1192. https://doi.org/10.3390/atmos11111192.

Southgate V. (1997). Schistosomiasis in the Senegal River basin: before and after the construction of the dams at Diama, Senegal and Manantali, Mali and future prospects. *Journal of Helminthology*, **71**(2), 125–132. doi: 10.1017/S0022149X00015790.

Tilmant A., Pina J., Salman M., Casarotto C., Ledbi F. and Pek E. (2020). Probabilistic trade-off assessment between competing and vulnerable water users – the case of the Senegal River basin. *Journal of Hydrology*, **587**, 124915. doi: 10.1016/J.JHYDROL.2020.124915.

Tractebel, OMVS (2013). Evaluation regionale strategique des options de developpement hydroelectrique et des ressources en eau dans le bassin du fleuve Senegal – Rapport d'evaluation regionale strategique – Volume 1.

Troussellier M., Got P., Bouvy M., M'Boup M., Arfi R., Lebihan F., Monfort P., Corbin D. and Bernard C. (2004). Water quality and health status of the Senegal River estuary. *Marine Pollution Bulletin*, **48**(9–10), 852–862. ISSN 0025-326X, https://doi.org/10.1016/j.marpolbul.2003.10.028.

UN (2003). Water for people, water for life: the United Nations world water development report; a joint report by the twenty-three UN agencies concerned with freshwater. UNESCO World Water Assessment Programme. ISBN: 978–92-3-103881-5, 92-3-103881-8, 1-57181-628-3, 978-89-8225-787-2 (kor).

UN (2021). UN list of Least Developed Countries [WWW Document]. URL https://unctad.org/en/Pages/ALDC/LeastDevelopedCountries/UN-list-of-Least-Developed-Countries.aspx (accessed 25 September 2020).

UNDP (2018). Human Development Indices and Indicators: 2018 Statistical Update. Available at: http://hdr.undp.org/en/content/human-development-indices-indicators-2018-statistical-update (last accessed July 2021).

United Nations (UN) (2015). Transforming our world: the 2030 Agenda for Sustainable Development [WWW Document].

World Health Organization (2009). Environmental Burden of Disease: Country Profiles. Geneva.

Assessing the WEFE Nexus and finding optimal solutions in the Mékrou transboundary river basin

C. Dondeynaz, M. Pastori, I. Ameztoy and C. Carmona-Moreno

Joint Research Centre, European Commission, Directorate D – Sustainable Resources, D2 – Water and Marine Resources Unit, Ispra (Va), Italy

ABSTRACT

The Mékrou is a transboundary river basin across Bénin, Burkina Faso, and Niger where the WEFE Nexus should be managed to ensure the sustainable development of this region. The management of the W Park ecosystems, the agriculture (both crop and livestock) production, the relatively poor level of access to basic services (water and energy) are to be improved altogether to alleviate poverty and push for sustainable development. In this project, two components interacted to reach two main objectives: (a) define a governance framework to allow to define a strategic and associated investment plan for the Mékrou River Basin and (b) develop an information system to support this strategic decision making. The institutional component adopted an iterative and participatory bottom-up approach to identify the key priorities of the Mékrou in 2015 and to discuss the strategic framework (CaSSE) options in 2017. From this CaSSE, an action (SDAGE) and an investment plan (PMPI) were discussed by all stakeholders before agreement. In support of the former, the scientific component designed and set up an information system called 'E-water' that integrates tools and analyses related to agricultural production, water resources, the value of the W Park ecosystems, and living conditions. The associated decision support tool is a multi-objective optimization in that it allows to provide optimal solutions according to several competing objectives to the decision maker. This paper will first describe both institutional and scientific processes and their interaction as well as the tools and methods used to encompass the breadth of the WEFE Nexus. The key achievements and examples of the analyses are then detailed before providing the lesson learnt and recommendations.

Keywords: WEFE Nexus, climate, agriculture, W park, Mékrou River Basin

CS6.1 INTRODUCTION

The Mékrou River Basin (Figure CS6.1) is a transboundary sub-basin of the Niger River Basin that covers an area of 10,635 km^2 or about 3% of the total Niger River Basin area. The Mékrou River Basin is shared

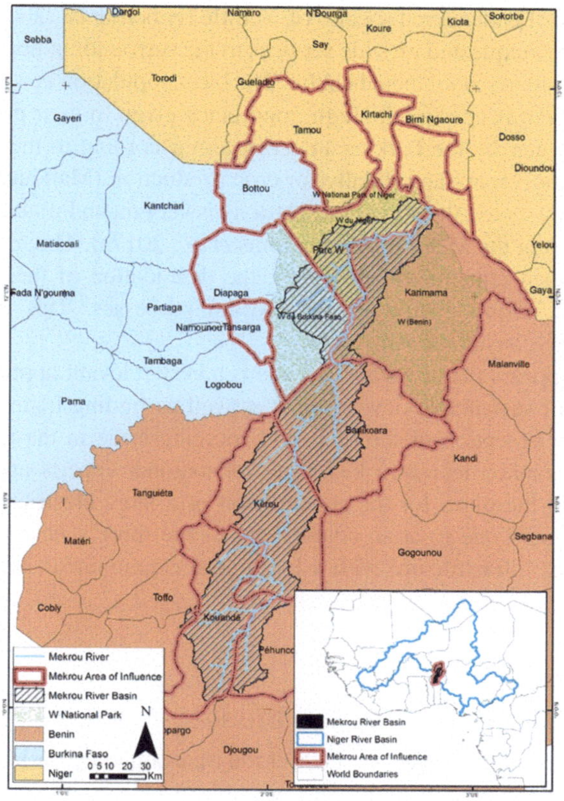

Figure CS6.1 Location of the Mékrou River Basin and area of influence.

among Benin (80%), Burkina Faso (10%), and Niger (10%). Agriculture is the key sector of the economy in the three riparian countries, and is also critical for poverty alleviation and food security. The arable land is mainly used for crop production and for raising cattle. The crops are rain-fed and what is produced are cereals (sorghum, maize, rice), cotton in Benin, cowpeas and yam/cassava. There also exists a small production of legumes. The Mékrou region is subject each year to the 'Grande Transhumance' of livestock according to an axis North–South from Nigerien-Burkinabe areas to Beninese areas (DED, 2006). In this configuration, the competition for water resources is mainly between domestic, crop production, and livestock demands. In addition, food security within the Mékrou River Basin is neither completely ensured nor homogeneous: 81% of the surveyed Beninese, 72% of the Burkinabe and only 55% of Nigerian declared that they could satisfy their family food needs during the last week prior to the 2016 survey. This situation is even less secure when looking back 12 months (2015) as less than 50% of surveyed households declared to have been able to fully cover their family food needs. This insecurity is more acute in Niger and Burkina Faso. The main reason indicated by far was the low harvest following a drought, while 8% had 'not enough money' (Markantonis *et al.*, 2017a).

In addition, there is an environmental water demand to consider, as the Mékrou River Basin includes a very important transboundary national park, the 'W Park'. The W Park belongs to the W-Arly-Pendjari (WAP) transboundary complex and is known to shelter the largest and most important continuum of terrestrial, semi-aquatic and aquatic ecosystems in the West African savannah (Amahowé *et al.*, 2013). The W Park is emblematic of transitional areas from Sahelian to savanna vegetation in West Africa, which hosts a high diversity of endangered species of wildlife (Hibert *et al.*, 2004; Michelot & Ouedraogo, 2009). These ecosystems are unique and provide services to the surrounding population. An environmental water availability is required to maintain them healthy. The local population is well aware of the value of the W Park and the Mékrou River as attested the high importance given to their preservation. The main uses from service of providing water by the Park as an ecosystem according to the surveyed families are: for livestock, for human, for biodiversity and, finally for crop production (Markantonis *et al.*, 2017a).

With regards to the Energy sector, the access to electricity is very much limited with an average of 20% of the population having access in the Mékrou (Markantonis *et al.*, 2017a). The main source of energy of the households living in the basin remains wood. Hence, the degradation of the forest, in particular in the upstream part of the river basin, leading to soil erosion processes. This issue was noted by local stakeholders in Benin during consultations in 2015.

From this overview, it appears that theWEFE Nexus is a very relevant approach for the Mékrou River Basin to define a strategy of sustainable development as well as finding trade-offs between the different uses of resources. With the exception of the fuelwood sourcing issue in the upstream part of the basin, this is despite energy was ranked less of a key issue for national experts and local stakeholders. This consideration was including the plans for the construction of a large electricity dam that were found to be of strong relevance for only one riparian country out of the three. These plans appear to have been replaced by a multi-purpose infrastructure with a hydropower component, the feasibility of which still needs to be assessed.

CS6.2 METHODOLOGY

CS6.2.1 Implementation approach of the information system NEXUS

The general approach adopted in the project of the Mékrou transboundary basin follows a number of steps: (a) review, processing, and integration of all the data available to assess the socio-economic and biophysical context of the Mékrou; (b) identification of the key issues and priorities for development by local stakeholders; (c) review of the relevant methods and tools available according to the outputs of the two previous steps; (d) further development and adaptation of selected tools/methods with local scientific and technical partners into an easy-to-use module (E-WATER); and finally, (e) running of models/analysis to test scenarios of development on the Mékrou River Basin as foreseen by the policy makers (i.e., in plans and strategies such as CaSSE and SDAGE[3]). The interaction between science and policy, including by local institutions, is crucial along this process. This is illustrated by the more detailed account of how the approach was implemented in practice as follows.

The data review required mobilizing the technical services of meteorology, hydrology, and agriculture of the three countries of the Mékrou as well as the W Park management. In addition, two field campaigns were conducted to obtain detailed data from the local population on their living conditions, the access to water resources and their different uses (agriculture, livestock, domestic, ecosystems…etc.) and the role of the park in securing that water supply. This data collection effort was carried out together with a literature review of technical and policy documentation to fill data gaps as much as possible. The main issues at this point were how to treat missing data, the time length of datasets and the low numbers of meteo/hydrological ground stations in the Mékrou.

In parallel with this review, several consultations of key stakeholders (local institutions, village assembly…etc.) in the form of dialogue forums were held in the three countries to discuss and list priorities first at sub-national scale and then at Mékrou River Basin (refer to Section 3.2). These

[3]CaSSE stands for Cadre Stratégique pour la Sécurité en Eaux, SDAGE for Schéma d'Aménagement et de Gestion des Eaux.

priorities were thematically grouped as follows: agricultural/crop production, rural development (in particular regarding livestock management), water resources management, and allocation between uses, climate variability (precipitation frequency and analysis of heatwaves), ecosystem protection including for tourism in the W Park, and overall socio-economic development in the Mékrou. Based on the data available, a review of the suitable methods and tools was made focusing on what could best fit the priorities identified by the stakeholders. Hereafter, the description of the tools selected, their advantages and their application to the Mékrou (Table CS6.1).

CS6.2.2 Application of selected tools to Mékrou River Basin

A challenge, as often in Africa or the developing world, is the data completeness and availability. This constraint has been overcome by first combining regional data, for instance with ground stations data (for precipitation/discharge) or results from the household survey (for instance in terms of food diet). The combinations of all data sources as well as the selection of appropriate methods are essential steps to ensure the reliability of the database and modelling. In addition, the methods selected are also coping with common low density of ground stations.

In the frame of the Mékrou project, SWAT and EPIC are combined and used to, first, assess climate change impacts on freshwater resources in the river basin using scenario CORDEX AFR-44 RCP 4.5, and; second, generate scenarios according to computed future water demand and land- use distribution to ensure crop production and livestock growth under demographic pressure. This is to identify hotspots or stresses in reaching the objectives set at the horizon 2030 by the SDAGE. The socio-economic data, EPIC and SWAT outputs feed the module of MOO to compute optimal solutions according to defined objectives.

In fact, in the Mékrou, the main objectives are the priorities identified by the stakeholders and are therefore multiple. In the MOO module, the objectives are related to fertilization, irrigation, livestock, revenue from crops production, and food demand. For example, an objective can be increasing the crop production, using fertilization and/or irrigation and/or crop distribution change and livestock. Another objective is to satisfy the domestic water demand (expected to increase by 3–4% per year) while preserving the other water demands. The MOO will identify optimal combination of the parameters that would ensure the objectives.

In the E-WATER, these tools are gathered together to allow a future user to run her/his simulations more easily, and to visualize and export specific results. This is an open-source software and tools are themselves freely available (avoiding software license costs).The scientific outputs were used to adjust or refine the CaSSE strategy developed by policy makers during the project. This followed an iterative process between the scientific and policy components (refer to Section 3.2).

CS6.3 KEY FINDINGS

The lesson learnt of the Mékrou project stem from two types of achievements: technical and of policy support.

CS6.3.1 Technical achievements in terms of WEFE Nexus assessment

As a key issue in the Mékrou, crop production can be simulated according to climate scenarios, distribution of crops (cropland areas) and management practices (mainly irrigation and fertilization). In the below example, the E-Water optimization tool allows to identify optimal strategies to efficiently use land resources combined with increased fertilization and irrigation strategies: optimal crop distribution and

Table CS6.1 Description of tools and methods used in E-WATER.

Name	Short description	Advantages	Reference
EPIC[1]	The Environmental Policy Integrated Climate (EPIC) model is a field-based cropping system model for assessing the effect of agricultural management practices on crop yield productivity, soil and water quality, and nutrient and pesticide flows and transformations	• Integration with GIS (Pastori *et al.*, 2011) • Extensive use in several contexts (Folberth *et al.*, 2012; Quiao *et al.*, 2018; Van der Velde *et al.*, 2014; Wriedt *et al.*, 2009).	Williams (1995)
FA	The frequency analysis (FA) of climate variables is a methodology using L-moments, and optionally a regionalization technique to analyse the frequency of extreme weather events, excess or deficit of climate variables, and estimate the time of return of such events.	• Accept both remote sensing and ground station data • Cope with low density of ground stations and short climate time series • Less sensitive to outliers	Hosking and Wallis (1997)
SPI	The Standardized Precipitation Index (SPI-n) is a statistical indicator comparing the total precipitation received at a particular location during a period of *n* months with the long-term rainfall distribution for the same period of time at that location.	• Applicable to all climatic regions • Easily readable	McKee *et al.* (1993, 1995)
Heat Wave Magnitude Index	The Heat Wave Magnitude Index is defined as the maximum magnitude of the heat waves in a year, where heat wave is the period $\geq$ 3 consecutive days with maximum temperature above the daily threshold for a year reference 30 period.	• Possible comparison across regions	Russo *et al.* (2014)

(*Continued*)

Table CS6.1 Description of tools and methods used in E-WATER (*Continued*).

Name	Short description	Advantages	Reference
SWAT	The Soil and Water Assessment Tool (SWAT)[2] is a process-based, semi-distributed, basin-scale model. It operates at daily time step and is designed to predict the impact of land use and management on water quantity and quality (water stream flow discharge, flows in the soil, sediments, and nutrients cycles) as affected by bio-physical conditions.	• Flexible structure • Cope with resolutions and quality of input data • Well documented model and widely used in many large basins around the world (Arnold *et al.*, 2012; Gassman *et al.*, 2007, 2014; Malagó *et al.*, 2017) • Considers climate scenarios	Arnold *et al.* (1998)
MOO	Statistical Multi-Objective Optimisation (MOO) techniques are used to identify combinations, named optimal solutions, according to multiple objectives in a potentially constrained environment.	• Multiple and flexible optimal solutions identified (Pastori *et al.*, 2017) • Decision makers are directly involved in the selection of the solution	Cortez (2014), Pareto (1971)

[1]EPIC description available here: https://epicapex.tamu.edu/epic/ last accessed 20/10/2017.
[2]SWAT documentation: http://swat.tamu.edu/documentation/ last accessed 13/10/2017.

fertilization/irrigation plans (which crop, how much and where) can be identified. The first scenario is based on the existing agricultural practices, the second one simulates the adoption of more intensive agriculture strategies and the third one is based on the combination of the second scenario and the possibility to change crop land use. The tool is directly addressing the food security and the satisfaction of local needs as one of the first set requirements (constraint) of the optimization. Concretely, for each food crop item, it is possible to estimate the missing quantity required to satisfy local diet demand (food item unfeasibility). In the Mékrou assessment, the total food unfeasibility is reduced by 61% with the second scenario and by 96% with the third scenario (Figure CS6.2).

The food security scenarios include African climate scenarios and can be cross-analysed with time return of extreme events such as precipitation deficits and heatwaves (Markantonis *et al.*, 2017b).

Another key issue is to ensure the satisfaction of water demands: current and future as set in the SDAGE. Similarly to the agricultural optimization, the e-water allows to find combination of water demands for irrigation, population and livestock according to objectives (i.e., increase the water available

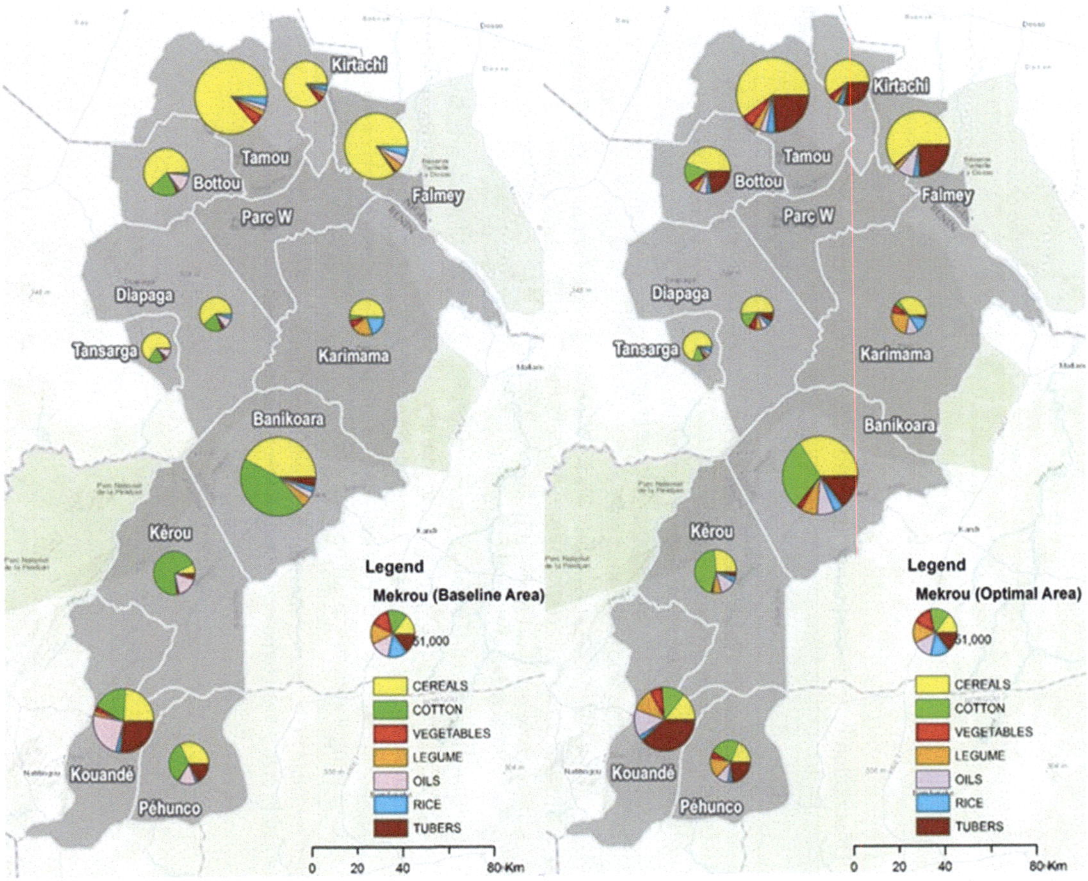

Figure CS6.2 E-water land use by region for different scenarios. Left: distribution of crop groups in the communes as defined for first scenario (cereals dominant in Niger and Burkina Faso; cotton in Benin). Right: distribution of crop groups for the third scenario (part of cereals replaced by tubers in Niger; significant reduction of cotton in Benin; increase in vegetables in Benin and Burkina Faso).

for irrigation by 10%) and constraints (i.e., the population water demand is satisfied at 100%), that minimize the exploitation of water resources available. The spatial repartition and the magnitude of the related effort are also computed.

CS6.3.2 Policy support

As a preliminary remark, the MEKROU project proceeded to a near-WEFE Nexus assessment of this river basin despite the fact that it was not defined to do so a priori. The need for this arouse from the local dialogues in each of the countries (in 2015–2016) where local stakeholders raised their expressed priorities for the water, agricultural/food, energy and environmental sectors that needed reconciling and optimization.

To ensure institutional support in this multi-sectorial context, the ministries and their technical services of meteorology, hydrology and agriculture were engaged into the whole process and the development of the E-water module. This systematic involvement has also some advantages in conjunction with the technical activities: (a) the application of methods and the results benefit from the local expertise who can detect errors and enrich the analyses or results, (b) the ownership of E-water tool is strengthened as it is tested and adapted by these experts, and (c) it provides a concrete dialogue space within which stakeholders from several countries can work together on a common product.

The E-Water tool and the scenarios that had been run with it fed the policy process necessary to define a strategic framework for action (SDAGE) at the horizon 2025 for the whole Mékrou River Basin. The SDAGE aimed at defining scenarios with priorities and quantifiable objectives for the development of the Mékrou that would be related to: transboundary governance, crop production, livestock, fishing, environmental preservation and conservation, access to water supply and sanitation, water resources management, and industrial/handcraft activities. The definition of the SDAGE resulted in an iterative process that took place between policy and science. First, four options of development for the Mékrou River Basin were defined. These have been simulated within the different models to adjust the figures and assess the feasibility of such scenarios. As an example, the objective of livestock growth was reviewed to a smaller number after estimating the forage (pasture/crop residue) and water required to allow such growth. After a multi-stakeholder dialogue with local representatives from the three countries that looked at these scientific inputs, a revised scenario for the basin has been consolidated and adjusted. At this stage, the outputs of regional simulations were also included in the discussion together with inputs from other stakeholders. This has led to the agreement on a joint SDAGE and a related action plan (Plan de Mesures et d'investissements).

CS6.4 DISCUSSIONS, CONCLUSIONS, AND RECOMMENDATIONS

The WEFE Nexus approach fits a reality of a local population where these four sectors are perceived as important in their daily life. During the project, the local stakeholders raised concerns regarding: (i) the management of pressures on resources (water, forests, etc.) for multiple purposes (crop production, livestock breeding, fuelwood for cooking, etc.) in a context of climate change and variability, (ii) the preservation of ecosystems in the W Park and, the associated benefits (pool of resources and tourism), (iii) the access to basic services, and (iv) the management/mitigation of floods and droughts.

One challenge was to integrate these local priorities into the transboundary scale in order to develop a regional development strategy for the river basin while keeping coherent with a national policy frame. The project followed an iterative bottom-up loop in this attempt. In particular, local dialogues with municipalities, user communities, etc. were initially held in each sub-national area. The identification of common issues and main specificities in each sub-area allowed the consolidation of a synthesis on the

Mékrou River Basin. Several scenarios and development objectives have been defined building up on this synthesis together with the review of the main appropriate strategies of the three Mékrou countries. These scenarios went back to the local level for discussions and were ranked by order of preference, before going through the transboundary dialogue platform for consolidation as Mékrou SDAGE. This method required time and resources for instance because using a not necessarily common consultative framework like the dialogue forums and platform implies more efforts to proceed with the regional synthesis. The regionally agreed SDAGE is a framework of objectives with a main axis for actions that is then detailed by sub-national areas and municipal level, giving the flexibility to address the local specificities identified.

In terms of scientific requirements, the WEFE Nexus assessment demands gathering data in all these different sectors. The main issue encountered in terms of data, as often in the developing world, was the short time series of data and the incomplete series. Methods to cope with this were applied to ensure robustness of the analyses. The combination of multi-source data like remote sensing regional datasets with the few ground station(s) and field campaign results is a robust solution. The E-Water integrates all the tools and analyses applied to the Mékrou River Basin into a user-friendly thematic interface (Figure CS6.3).

The E-Water open-source software was a necessary condition to ensure its use by all stakeholders and the flexibility to replicate them to other river basins. The technical services and scientific experts (i.e., from

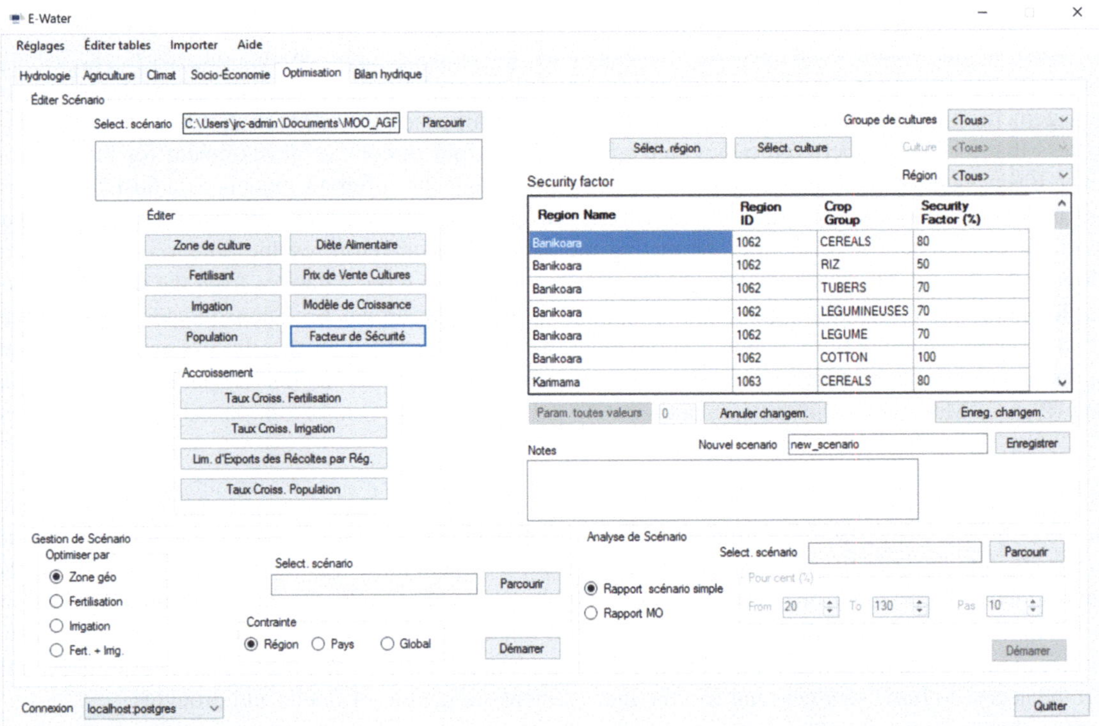

Figure CS6.3 E-water interface – Generation of WEFE Nexus scenarios. *Hydrology*: SWAT scenarios; *Agriculture*: Agricultural scenarios; *Climate*: climate variability analyses and maps; *Socio-economic*: household survey results; *Optimization agri:* Multi-objective optimization of agriculture production; *Optimization water:* optimization of water demands vs water availability.

research institutes or universities) of the three countries were trained and their feedbacks were integrated during the iterative development cycle of the module, thus ensuring its ownership by the local stakeholders.

The following recommendations can be drawn from the Mekrou case:

- Delivering scientifically sound information on a river basin can and should contribute to the dialogue among sectors and countries necessary to address NEXUS assessment and strategy. This can be done through technical workshops with experts from ministries who work in this field on a daily basis. This is important to ensure as much as possible that the technical experts of ministries who participate in the scientific activities are the same. A clear profile of the competences required to actively participate in this type of workshop may help the nomination of the appropriate participant by the relevant ministry.
- The inclusion of appropriate universities and research institutions reinforces the participation of the local expert base and fosters the collaboration between the institutional (i.e., technical service of hydrology or agriculture) and the research side. This inter-sectorial, inter-institutional link developed during a nexus assessment will potentially continue afterwards because people learn to know/trust each other while working on a concrete activity.
- The Information/IT tools have to be open source, to ensure wide use and replicability.
- Developing the information system E-WATER and the strategic planning platform (i.e., SDAGE and PMPI) in parallel with the stakeholders is key in order to ensure maximum synergy. To this effect, the interaction between the two has to be iterative: potential scenarios/objectives of development are tested by available tools. The outputs are then sent back to policy makers and stakeholders to adjust through revision of the scenarios/objectives. The time needed for this interactive process requirement should not be neglected.

REFERENCES

Amahowé I. O., Houessou L. G., Ashanti S. and Tehou A. C. (2013). Transboundary protected areas management: experiences from W-Arly-Pendjari Parks in West Africa. *Parks*, **19**(2), 95–015.

Arnold J. G., Srinivasan R., Muttiah R. S. and Williams J. R. (1998). Large area hydrologic modeling and assessment part I: model development. *Journal of the American Water Resources Association*, **34**, 73–89. doi: 10.1111/j.1752-1688.1998.tb05961.x.

Arnold J. G., Moriasi D. N., Gassman P. W., Abbaspour K. C., White M. J., Srinivasan R., Santhi C., Harmel R. D., van Griensven A., Van Liew M. W., Kannan N. and Jha M. K. (2012). SWAT: model use, calibration, and validation. *Transactions of the ASABE*, **55**, 1491–1508. doi: 10.13031/2013.42256.

Cortez P. (2014). Modern Optimization with R Springer. Springer International Publishing, Switzerland, ISBN 978-3-319-08263-9.

Deutscher Entwicklungsdienst (DED) (2006). Les conflicts liés à la transhumance transfrontalière entre le Niger, le Burkina Faso et le Benin, 79 pp. http://www.peaceresources.net/files/docs/publications/DEDexpert_Transhumanzstudie.pdf (accessed 27 October 2017).

Folberth C., Gaiser T., Abbaspour K. C., Schulin R. and Yang H. (2012). Regionalization of a large-scale crop growth model for sub-Saharan Africa: model setup, evaluation, and estimation of maize yields. *Agriculture, Ecosystems & Environment*, **151**, 21–33. doi: 10.1016/j.agee.2012.01.026.

Gassman P. W., Reyes M. R., Green C. H. and Arnold J. G. (2007). The soil and water assessment tool: historical development, applications, and future research directions. *Transactions of the ASABE*, **50**, 1211–1250. doi: 10.13031/2013.23637.

Gassman P. W., Sadeghi A. M. and Srinivasan R. (2014). Applications of the SWAT model special section: overview and insights. *Journal of Environmental Quality*, **43**, 1. doi: 10.2134/jeq2013.11.0466.

Hibert F., De Visscher M. N. and Alleaume S. (2004). The wild ungulate community in the Niger W Regional Park. Antelope survey update (eds.), pp. 31–35.

Hosking J. R. M. and Wallis J. R. (1997). Regional Frequency Analysis: An Approach Based on L-moments. Cambridge University Press, Cambridge, UK.

Malagó A., Bouraoui F., Vigiak O., Grizzetti B. and Pastori M. (2017). Modelling water and nutrient fluxes in the Danube River Basin with SWAT. *Science of the Total Environment*, **603**, 196–218. doi: 10.1016/j.scitotenv.2017.05.242.

Markantonis V., Dondeynaz C., N'Tcha M'po Y., Sanon K., Ameztoy-Aramendi I., Alhassane A., Degnide A. M. and Carmona-Moreno C. (2017a). « Bassin versant de la Mékrou: Enquête sur les ménages. Technical report, Publication Union Européenne, 90 pp.

Markantonis V., Farinosi F., Dondeynaz C., Ameztoy I., Pastori M., Marletta L., Ali A. and Carmona Moreno C. (2017b). Assessing floods and droughts in the Mékrou River Basin (West Africa): A combined household survey and climatic trends analysis approach. *Natural Hazards and Earth System Sciences Discussion*, 1–55. https://doi.org/10.5194/nhess-2017-195.

McKee T. B., Doesken N. J. and Kleist J. (1993). The Relationship of Drought Frequency and Duration of Time Scales. Eighth Conference on Applied Climatology, American Meteorological Society, Jan 17–23, 1993, Anaheim, CA, pp. 179–186.

McKee T. B., Doesken N. J. and Kleist J. (1995). Drought Monitoring with Multiple Time Scales. Nineth Conference on Applied Climatology, American Meteorological Society, Jan 15–20, 1995, Dallas, TX, pp. 233–236.

Michelot A. and Ouedraogo B. (2009). Transboundary Protected Areas: Legal Framework for the W Transboundary Biosphere Reserve (Benin, Burkina Faso, Niger). Environmental Policy and Law Paper (EPLP) No. 81, Gland, Switzerland, International Union for Conservation of Nature (IUCN), pp. 1–33.

Pareto V. (1971). Manual of Political Economy. Kelley Publ., New York.

Pastori M., Bouraoui F., Aloe A., Bidoglio G. and European Commission. Joint Research Centre. Institute for Environment and Sustainability (2011). GISEPIC AFRICA: a modeling tool for assessing impacts of nutrient and water use in African agriculture (database, model and GIS system development and testing). Publications Office.

Pastori M., Udías A., Bouraoui F. and Bidoglio G. (2017). A multi-objective approach to evaluate the economic and environmental impacts of alternative water and nutrient management strategies in Africa. *Journal of Environmental Informatics*, **29**, 16–28. doi: 10.3808/jei.201500313.

Quiao J., Yu D. and Wu J. (2018). How do climatic and management factors affect agricultural ecosystem services? A case study in the agro-pastoral transitional zone of northern China. *Science of the Total Environment*, **613–614**, 314–323. doi: 10.1016/J.SCITOTENV.2017.08.264.

Russo S., Dosio A., Graversen R. G., Sillmann J., Carrao H., Dunbar M. B., Singleton A., Montagna P., Barbola P. and Vogt J. V. (2014). Magnitude of extreme heat waves in present climate and their projection in a warming world. *Journal of Geophysical Research. Atmospheres: JGR*, **119**, 12,500–12,512. doi: 10.1002/2014JD022098

Van der Velde M., Folberth C., Balkovič J., Ciais P., Fritz S., Janssens I. A., Obersteiner M., See L., Skalský R., Xiong W. and Peñuelas J. (2014). African crop yield reductions due to increasingly unbalanced nitrogen and phosphorus consumption. *Global Change Biology*, **20**, 1278–1288. doi: 10.1111/gcb.12481.

Williams J. R. (1995). The EPIC model. In: Computer Models of Watershed Hydrology, Vijay P. Singh (ed.). Water Resources Publications, Highlands Ranch, CO, USA.

Wriedt G., Van der Velde M., Aloe A., Bouraoui F. F., der Velde M., Aloe A. and Bouraoui F. F. (2009). Estimating irrigation water requirements in Europe. *Journal of Hydrology*, **373**, 527–544.

Water–energy–food–ecosystem Nexus assessment in the Blue Nile Basin in Sudan

M. Basheer and M. A. Gamal

Water Research Center, University of Khartoum, Khartoum, Sudan

ABSTRACT

This study analyses past and future scenarios for the WEFE Nexus in the Blue Nile Basin in Sudan. Water availability, hydropower generation, irrigation water supply, and environmental flows are the components considered in the current assessment. A calibrated daily rainfall–runoff and water allocation model was used to quantify the four nexus components and their interlinkages. The model includes three storage reservoirs (Grand Ethiopian Renaissance Dam, Roseires Dam, and Sennar Dam), seven inflow nodes, five irrigation demand nodes, evaporation losses from reservoirs, return flows from irrigation schemes, and transmission losses from river reaches. Due to a scarcity of ground rainfall data in the study area, a pixel-to-point evaluation was conducted for four satellite-based rainfall products, and the best-performing one was used as a boundary condition for the rainfall–runoff component of the model. The model was used to assess the historical and future association of seven WEFE Nexus indicators. The results show a historical association between environmental flow supply, irrigation water supply, and water availability. The heightening of the Roseires Dam in 2013 affected most of the nexus indicators. The results reveal that the steady-state operation of the Grand Ethiopian Renaissance Dam will positively affect irrigation water supply, hydropower generation, and environmental flows in the Blue Nile Basin in Sudan.

Keywords: irrigation, hydropower, environmental flow, water allocation, WEFE Nexus, Blue Nile

CS7.1 INTRODUCTION

It is widely acknowledged that overcoming the challenge of resource scarcity requires adopting integrated management approaches, as opposed to fragmented approaches, to achieve long-term WEFE sustainability (Hoff, 2011). The challenge of resource scarcity can be associated with two issues the world is going through. On the one hand, the rapidly swelling urbanization, the growth of the middle-income class, and the changes in lifestyles are placing stress on resources. In 1950, around one-third of the world's population lived in cities, while in 2000 every one in two people was a city dweller. The proportion of the urban population is projected to increase to two-thirds by 2030, which indicates a high growth rate of the middle-income group (UNDP, 2015). On the other hand, it is projected that the global population will grow to 8.5 billion by 2030, 9.7 billion by 2050, and 11.2 billion by 2100 (UN, 2015a). In 2015, about 800 million people were living in extreme poverty and suffered from hunger, approximately 1.2 billion people had income lower than 1.25 USD per day (Bhattacharyya *et al.*, 2015), and above 160 million children under age 5 had a low height for their age attributable to insufficient food (UN, 2015b). Water scarcity is affecting more than 40% of the world population. These effects will be amplified by climate change, which could expose around 250 million people to water stress in Africa alone (UNDP, 2015).

The WEFE Nexus theoretical framework illustrates a need to advance our understanding of the interactions between water use and management, energy production, food production, and environmental requirements. This knowledge is pivotal to circumvent future supply deficiencies that would hinder development and to ensure sustainable access to water, energy, and food while maintaining the environment in acceptable status. In recent years, a considerable amount of literature investigated the WEF Nexus in transboundary river basins. Bazilian *et al.* (2011) studied the interlinkages of WEF from a developing country perspective and argued that holistic treatment of the three resources in the context of transboundary river basins improves allocation, enhances economic efficiency, and minimizes negative externalities. Nevertheless, they found that tools and expertise are not yet available to bring the WEF Nexus approach into practice. Keskinen *et al.* (2015) applied the WEF Nexus approach to the Mekong River Basin and found that water and food security are likely to be altered by hydropower development. Keskinen *et al.* (2015) and Strasser *et al.* (2016) compared the Integrated Water Resources Management and the WEF Nexus approaches and concluded that the latter treats the water, energy, and food sectors in an equal manner. However, they found that some aspects, such as livelihoods, climate change, and the environment, are not explicitly considered in the WEF Nexus approach (Keskinen *et al.*, 2015). Kibaroglu and Gürsoy (2015) studied the evolution of transboundary WEF management policies in the Euphrates–Tigris River Basin and their impacts on cooperation between riparian countries. They found that the compound nature of pressures and drivers in the Euphrates–Tigris River Basin necessitates adopting a nexus approach to reach a win-win situation between riparian countries. Pittock *et al.* (2016) developed a comprehensive WEF Nexus framework for the Mekong River Basin that shows the interplay of WEF Nexus variables. Strasser *et al.* (2016) proposed a methodology to assess WEFE Nexus in transboundary river basins and presented results for the Alazani/Ganykh, the Sava, and the Syr Darya transboundary river basins.

Assessing the WEFE Nexus is often carried out by separate disconnected institutional entities. For instance, water management institutions often treat food and energy production as end-users; food and agricultural institutions see water and energy as production inputs; energy institutions treat water as an input resource (Howells *et al.*, 2013). The need for the WEFE Nexus approach originated from the growing scarcity, recent supply crises, and failures of individualism in sectorial management (Al-Saidi & Elagib, 2017). While the integrated management approach of water, food, energy, and the environment is relatively new (started in the 2000s), calls for water–food, water–energy, and food–energy nexus approaches dates back to programmes in the 1980s by the United Nations University (Bhattacharyya *et al.*, 2015).

The WEFE Nexus approach did not have much attention in Africa compared to other regions in the world. Endo *et al.* (2017) review several water, energy, food, and climate-related studies on Africa to assess their nexus orientation. They found that most of the studies on Africa did not include all nexus sides. Some nexus research has been recently conducted for the Nile Basin. Basheer and Elagib (2018a) examined the water–energy nexus for the White Nile and the Jebel Aulia Dam. They introduced the water–energy productivity, which is defined as the amount of water lost to evaporation from a reservoir for each unit of hydro-electricity generation. Basheer *et al.* (2018) explored the impact of transboundary cooperation in the Blue Nile Basin on the water–energy–food nexus. They found that a higher level of cooperation increases basin-wide benefits. Elagib *et al.* (2019) investigated the urban water–energy–food nexus in Khartoum State at the confluence of the Blue Nile and the White Nile. They found a strong relationship between hydrological phenomena (such as flood and drought) and the resource nexus. Stamou and Rutschmann (2018) analysed the trade-offs and synergies between hydropower generation and irrigation water supply in the Upper Blue Nile Basin using the parameterization–simulation–optimization method.

This study assesses the WEFE Nexus in the Blue Nile Basin in Sudan. The Blue Nile is the largest tributary of the Nile River in terms of annual flow contribution. We develop a descriptive framework for the interplay of water, energy, food, and ecosystem resources in the Blue Nile Basin in Sudan in the context of river water availability, hydropower generation, irrigation water supply, and environmental flows. We develop, calibrate, and validate water allocation and hydrological model for the Blue Nile in Sudan using RiverWare and HEC-HMS. The model includes three storage reservoirs (Grand Ethiopian Renaissance Dam, Roseires Dam, and Sennar Dam), seven inflow nodes, five irrigation demand nodes, evaporation losses from reservoirs, return flows from irrigation schemes, and transmission losses from river reaches. The model was used to assess the historical and future association of seven WEFE Nexus indicators, which are annual river flow, annual energy generation, evaporation losses, water–energy productivity, irrigation water supply shortage, risk of irrigation water supply shortage, and risk of environmental flow violation. Due to a scarcity of ground rainfall data in the Blue Nile Basin in Sudan, a pixel-to-point evaluation was conducted for four satellite-based rainfall products, and the best-performing one was used as a boundary condition for the rainfall–runoff component of the model. The evaluated satellite rainfall products include the African Rainfall Climatology Version 2 (ARC2.0), the Tropical Applications of Meteorology Using Satellite Data and Ground-Based Observations version 2 (TAMSAT2), the Precipitation Estimation from Remotely Sensed Information Using Artificial Neural Networks–Climate Data Record (PERSIANN-CDR), and the Climate Hazards group Infrared Precipitation with Stations version 2.0 (CHIRPS 2.0).

CS7.2 STUDY AREA

The study area extends over the Blue Nile reach from Grand Ethiopian Renaissance Dam (GERD) to the confluence point of the Blue Nile and the White Nile (see Figure CS7.1), including all major water inflows, abstractions, and infrastructures.

The Sennar Dam became operational in 1925 to supply 126,000 ha of cotton in the Gezira scheme with irrigation water by gravity from headworks located within the dam on the left bank of the Blue Nile. In 1962 two 7.5 MW turbines were installed in a power station on the west side of the dam to utilize the downstream flow for hydropower generation (MoIHES, 1977). Implementing the Managil extension of the Gezira Scheme led the Sudanese government in the year 1925 to investigate a proposal for the construction of a dam with a capacity of at least 1.0 BCM near the Roseires Town. Two years later, the location of the dam was confirmed. The provision of storage larger than 1.0 BCM became increasingly important, especially after the completion of the Managil extension. Therefore, in 1955 the Sudanese government appointed the firms of Sir Alexander Gibb & Partners and Coyne et Bellier of Paris to conduct a joint study on the consequences of constructing a larger dam at Roseires than the one proposed earlier in 1925. The two firms suggested a design for a dam that would be constructed in two stages (MoIHPS, 1966). The first stage of the dam was completed in 1966, followed by an attempt to construct the second stage in the 1990s, which stopped because of the economic situation of Sudan at that time (Roseires Dam Heightening Unit, 2005). The heightening of the dam started again in May 2009 and was finished

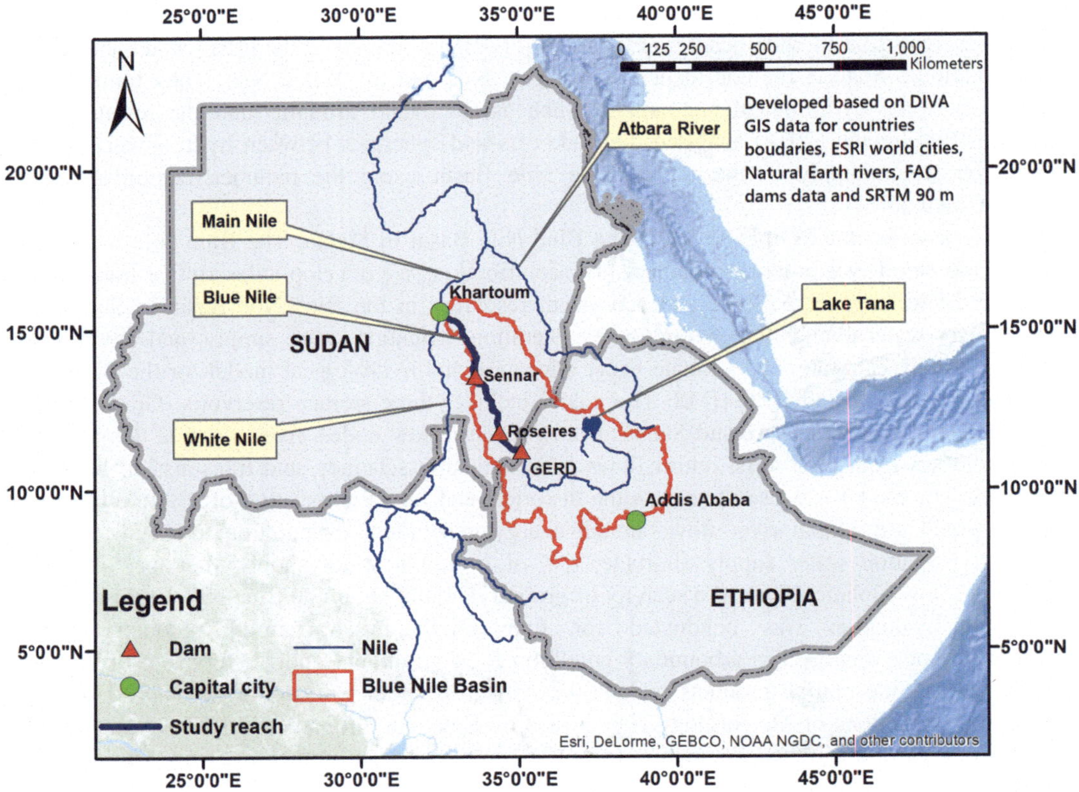

Figure CS7.1 General features of the Blue Nile Basin downstream the GERD. The names and boundaries shown and the designations used on this map do not imply official endorsement or acceptance by the European Commission or the United Nations.

in January 2013 (DIU, 2016). The water stored in the reservoirs of the Roseires and Sennar dams is essential for irrigated agriculture schemes because the flow of the Blue Nile is seasonally variable. In addition to providing agricultural water, Roseires and Sennar dams serve in hydropower production.

Large-scale agricultural development in the study area began in 1925 with the commissioning of the Gezira Scheme. During the late 1950s and 1960s, the Managil extension was constructed, which more than doubled the total area of the Gezira. The total area of Gezira and Managil schemes amounts to around 840,000 Ha. The Gezira and Managil remain the only gravity-fed scheme based on the Blue Nile in Sudan and represent more than 50% of irrigated agriculture in Sudan. Weighty development in pumping irrigation from the Blue Nile took place in the 19th century in reaction to the 1950s increase in cotton prices. The Blue Nile pumping developments include the construction of the Gunied, Rahad Phase1, and Suki Schemes, which took place during the late 1960s in addition to North West Sennar Sugar scheme in the 1970s (IWMI, 2012; MoIHES, 1977).

On 2 April 2011, the Ethiopian government announced the start of the currently under construction Grand Ethiopian Renaissance Dam (GERD) with a power capacity of 5150 MW, ranked as the largest in Africa and the tenth-largest globally. The GERD is located on the Blue Nile River 20 kilometers upstream of the Sudanese Ethiopian border.

CS7.3 METHODS

CS7.3.1 WEFE Nexus framework

Dams represent a clear example of the WEFE Nexus because they are often multipurpose and serve more than one sector. Figure CS7.2 shows a theoretical framework for the WEFE Nexus in the study area. The primary purpose of storage dams in the Blue Nile Basin in Sudan is to provide water for irrigation. This purpose sometimes contradicts with hydropower production depending on the location of the irrigation intakes with respect to turbines. Although dams can potentially provide benefits through storage, the larger the storage, the more water is lost to the atmosphere through evaporation. Evaporated water leaves the river system with no chance to be recaptured. The ecosystem is also often affected by dams. Usually, minimum environmental flows are reserved to ensure that the ecosystem is conserved. In principle, environmental flows conflict with irrigation water supply and hydropower generation since it reduces reservoir storage. However, in some cases, it positively affects hydropower production since the minimum environmental flow can be passed through hydro turbines to generate electricity (e.g., Sennar dam).

CS7.3.2 WEFE Nexus modelling

In this study, a daily water allocation and rainfall–runoff model was developed for the study area. The model was calibrated and validated over the 1984–2016 period. Figure CS7.3 shows a schematic of the model developed herein. The model includes three storage dams (GERD, Roseires Dam, and Sennar Dam),

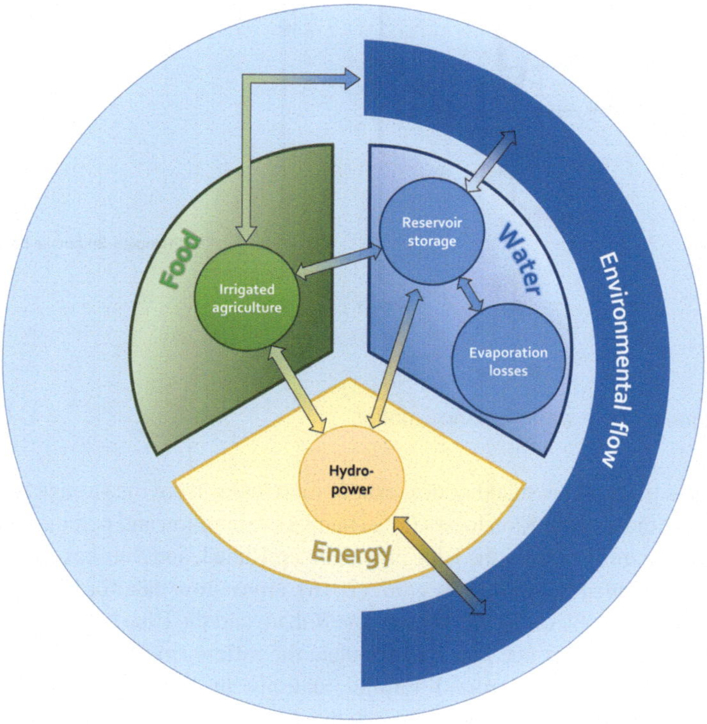

Figure CS7.2 Interconnections of water, energy, food, and ecosystem considered in this study.

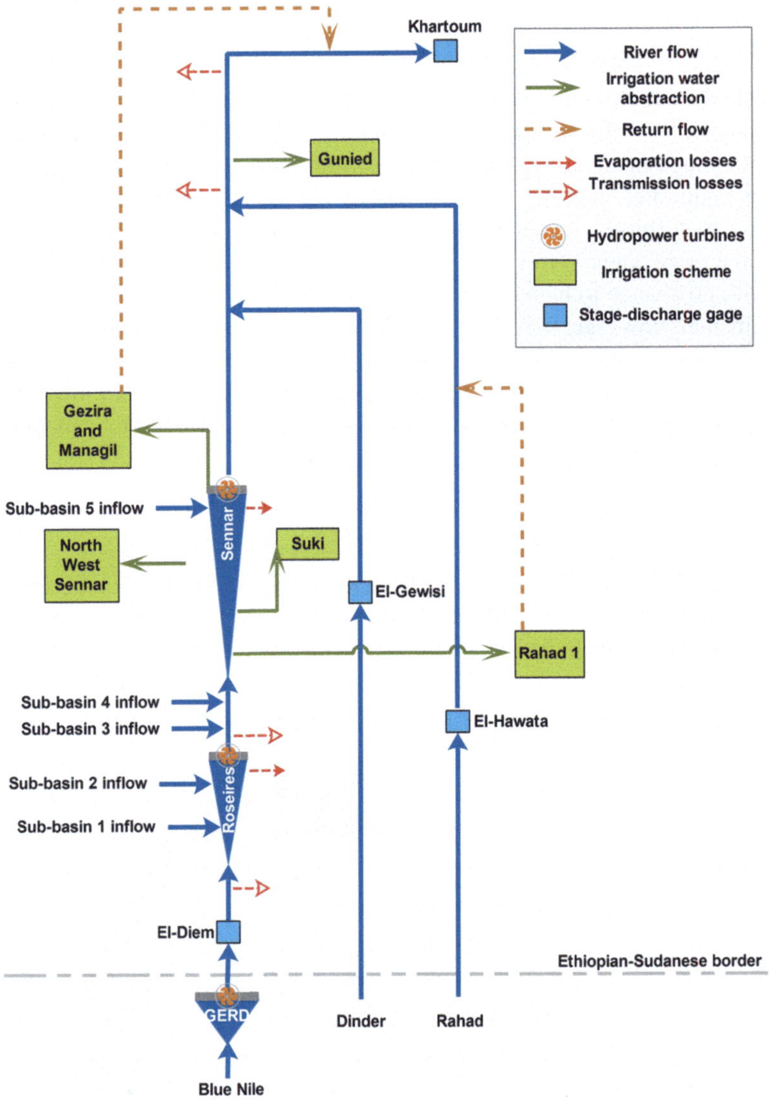

Figure CS7.3 Schematic of the water balance model.

seven inflow nodes, five irrigation demand nodes, evaporation losses from the storage reservoirs, return flows from irrigation schemes, and transmission losses (i.e., channel evaporation and percolation) from river reaches. The model is driven by the inflows from the Blue Nile, Dinder, Rahad, and Sub-basins 1–5 and dam operating rules. These rules were obtained from Basheer *et al.* (2018). River flow data for El-Gewisi, EL-Hawata, and El-Diem gauges were used as inflows for the Dinder, the Rahad, and the Blue Nile, respectively. The rainfall–runoff component of the model has been used to simulate the inflow from sub-basins 1–5 because they are ungauged. Due to the scarcity of ground rainfall stations in the study area, the performance of satellite-based rainfall products was evaluated, and the best performing one was used as a boundary condition to the rainfall–runoff component of the model (see Section 3.3).

Water allocation was simulated using RiverWare, a river and reservoir simulation software developed by the University of Colorado Boulder (Zagona *et al.*, 2001). RiverWare is capable of simulating hydraulic and hydrologic processes of reservoirs, river reaches, diversions, canals, abstractions, groundwater interaction, hydropower production, water ownership, and water accounting transactions. The object-oriented approach of RiverWare allows the user to create a network of objects, link them, populate each one with data, and select the appropriate physical process. RiverWare's rule-based simulation enables simulating operating policies using logical statements rather than explicitly specified input values for operations. HEC-HMS was used to simulate rainfall–runoff in the study area. HEC-HMS, developed by the Hydrologic Engineering Centre, is a freely accessible numerical model (computer program) that includes a variety of methods to simulate rainfall–runoff of dendritic watershed systems. HEC-HMS simulates watershed precipitation and evaporation, runoff volume, direct runoff, baseflow, and channel flow (HEC, 2008). R, an open-source programming language, was used in evaluating the performance of satellite-based rainfall products. R provides several packages and functions for downloading remote sensing data, extracting pixel values, and calculating the average of pixels within sub-basins (R Core Team, 2015).

CS7.3.3 Evaluation of satellite-based rainfall products

Due to the limited number of rainfall gauges in the study area, four satellite-based rainfall products have been evaluated, and the best performing one was used as a boundary condition to model rainfall–runoff in the study area. The evaluated satellite-based rainfall products include the African Rainfall Climatology Version 2 (ARC2.0; Novella and Thiaw, 2013), Tropical Applications of Meteorology Using Satellite Data and Ground-Based Observations version 2 (TAMSAT2; Maidment *et al.*, 2014), Precipitation Estimation from Remotely Sensed Information Using Artificial Neural Networks–Climate Data Record (PERSIANN-CDR; Ashouri *et al.*, 2015), and Climate Hazards group Infrared Precipitation with Stations version 2.0 (CHIRPS 2.0; Funk *et al.*, 2014).

To measure the difference between satellite estimates and ground observations, a pixel-to-point evaluation was conducted for the satellite products at the locations of five ground rainfall stations using the available measured data (1999–2009). Figure CS7.4 shows the locations of the five ground stations. Six performance metrics were used to conduct the evaluation. Those metrics can be categorized into two groups: (1) error metrics that include the root mean square error (RMSE; Chai and Draxler, 2014), the mean bias error (MBE; Legates and McCabe, 1999), and the coefficient of determination (R2; Legates and McCabe, 1999) (2) categorical metrics that include the probability of detection (POD; Toté *et al.*, 2015), the false alarm ratio (FAR; Diem *et al.*, 2014), and the equitable threat score (ETS; Ebert *et al.*, 2007).

To draw an overall conclusion on the best-performing product based on all performance metrics, the overall unified metric (OUM) was calculated for each of the evaluated satellite-based rainfall products. OUM is a performance metric developed by Basheer and Elagib (2018b). Equations (CS7.1) and (CS7.2) show the calculation procedure of OUM. High OUM values indicate poor performance and vice versa. We refer the reader to Basheer and Elagib (2018b) for further information.

$$\mathrm{UM}_{rj} = \sum_{i=1}^{p} R_{rji} \tag{CS7.1}$$

$$\mathrm{OUM}_{r} = \sum_{j=1}^{e} \mathrm{UM}_{rj} \tag{CS7.2}$$

where UM_{rj} is the unified metric of the rainfall product r at the station j, p is the number of aggregated performance metrics, R_{rji} is the performance ranking of the rainfall product r at the station j based on the

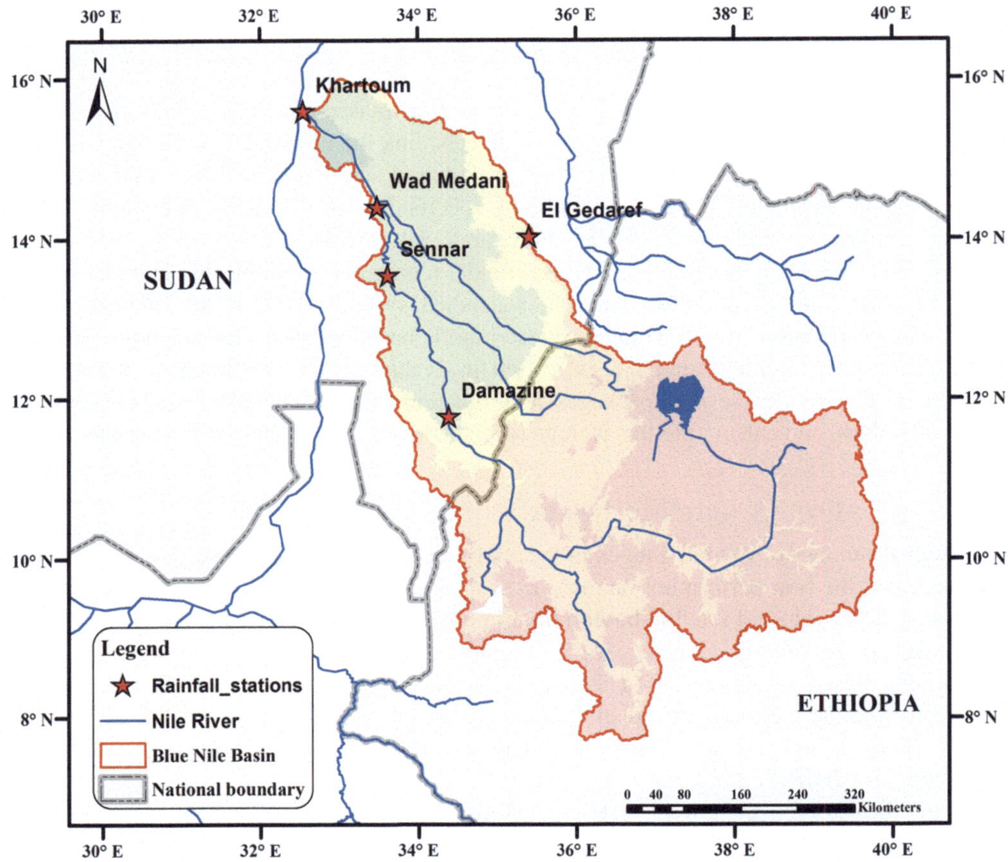

Figure CS7.4 Rainfall stations used in this study. The names and boundaries shown and the designations used on this map do not imply official endorsement or acceptance by the European Commission or the United Nations.

performance metric i, OUM_r is the overall unified metric of the rainfall product r, e is the number of stations, and UM_{rJ} is the unified metric of the rainfall product r at the station j.

CS7.3.4 Simulation scenarios

In this study, 34 simulation scenarios were examined. The examined scenarios comprise of a historic baseline scenario for the 1984–2016 period and a scenario with the GERD (in steady-state operation) on the river system. The latter scenario was examined across 33 hydrologic sequences (each 33 years long). The hydrologic sequences were developed using the index-sequential method (Kendall & Dracup, 1991; Ouarda *et al.*, 1997). In the 33 scenarios that include the GERD in full operation, the operation of the Roseires and Sennar Dams was modified to keep them at their full supply level. This modification has been recommended by several studies (Basheer *et al.*, 2018; Wheeler *et al.*, 2018, 2016). In this study, the steady-state operation of the GERD was assumed to target a power rate of 1400 MW. Wheeler *et al.* (2018) found that this power rate would maximize the firm annual energy generation of the dam.

CS7.4 RESULTS AND DISCUSSION

CS7.4.1 Performance of satellite-based rainfall products

Figure CS7.5 shows the performance metrics of the four satellite-based rainfall products at five locations. It is evident in the figure that the ARC2 has the best performance in terms of ETS, R2, RMSE, and FAR at all locations compared to the other rainfall products. ARC2 has the second-best performance in terms of POD and MBE.

Figure CS7.6 shows the overall unified metric of the four satellite-based rainfall products. The figure shows that ARC2 outperformed the other satellite-based rainfall products at all evaluation locations. Therefore, ARC2 has been used as a boundary condition to model rainfall–runoff in the study area.

CS7.4.2 Model performance

Table CS7.1 shows the performance metrics and performance ranking of the model at the Roseires and Sennar dams and the Khartoum Gage based on the recommendations of Stern *et al.* (2016) on performance ranking of hydrological models. The model accurately captured the inter- and intra-annual behaviour of the Blue Nile. The high R^2 values show that the variation in the simulated flow could explain a large portion of the variation in the observed flow. Generally, the model showed better performance in the calibration period than in the validation period. This is because there is uncertainty around the operation of the Roseires Dam since the heightening of the dam in 2013.

CS7.4.3 WEFE Nexus assessment

Figure CS7.7 shows the probability of exceedance of the annual flow of the Blue Nile at El-diem with and without the GERD. The figure shows that the steady-state operation of the GERD would reduce the inter-annual variability of the flow, as explained by a reduction in the maximum annual flow and an increase in the minimum annual flow compared to the baseline. The decrease in variability would positively affect water availability in the Lower Blue Nile Basin. However, it implies a negative impact on recession agriculture along the Blue Nile.

In this study, we used the annual energy generation and the annual water-energy productivity (WEP) as indicators for hydropower generation in the study area. The WEP is a water–energy nexus indicator developed by Basheer and Elagib (2018a) and is defined as the amount of energy produced per unit of water lost in the process. Figure CS7.8 depicts the WEFE Nexus indicators of the Blue Nile in Sudan. The colours in Figure CS7.8(a) distinguish the years with high and low water-energy productivity. The figure shows a historical annual energy generation in the study area in the range of around 1450–2410 GWh. The WEP took a range of 1775–2968 GWh/BCM from 1984 to 2016. Figure CS7.8(a) shows that most of the years with high annual energy generation and low WEP are after 2012. The heightening of the Roseires Dam in 2013 is the reason behind this behaviour. This heightening increased both energy generation and reservoir evaporation, with a higher increase in the latter than in the former resulting in a decrease in the WEP. The WEP can be increased for the Roseires Dam by utilizing the increase in hydropower potential that resulted from the heightening. Hydropower turbines with a high capacity could be installed in the Roseires Dam to achieve that. Operating the reservoir at lower level could also be used to increase the WEP; however, this will result in water supply shortages (SS) bearing in mind that hydropower generation is not the primary purpose of the dam.

The analysis of hydro-energy generation and reservoir evaporation in Sudan during the steady-state operation of the GERD reveals the following with all hydrologic conditions: total annual energy generation from the Roseires and Sennar Dams of around 2560 GWh, total annual evaporation from the

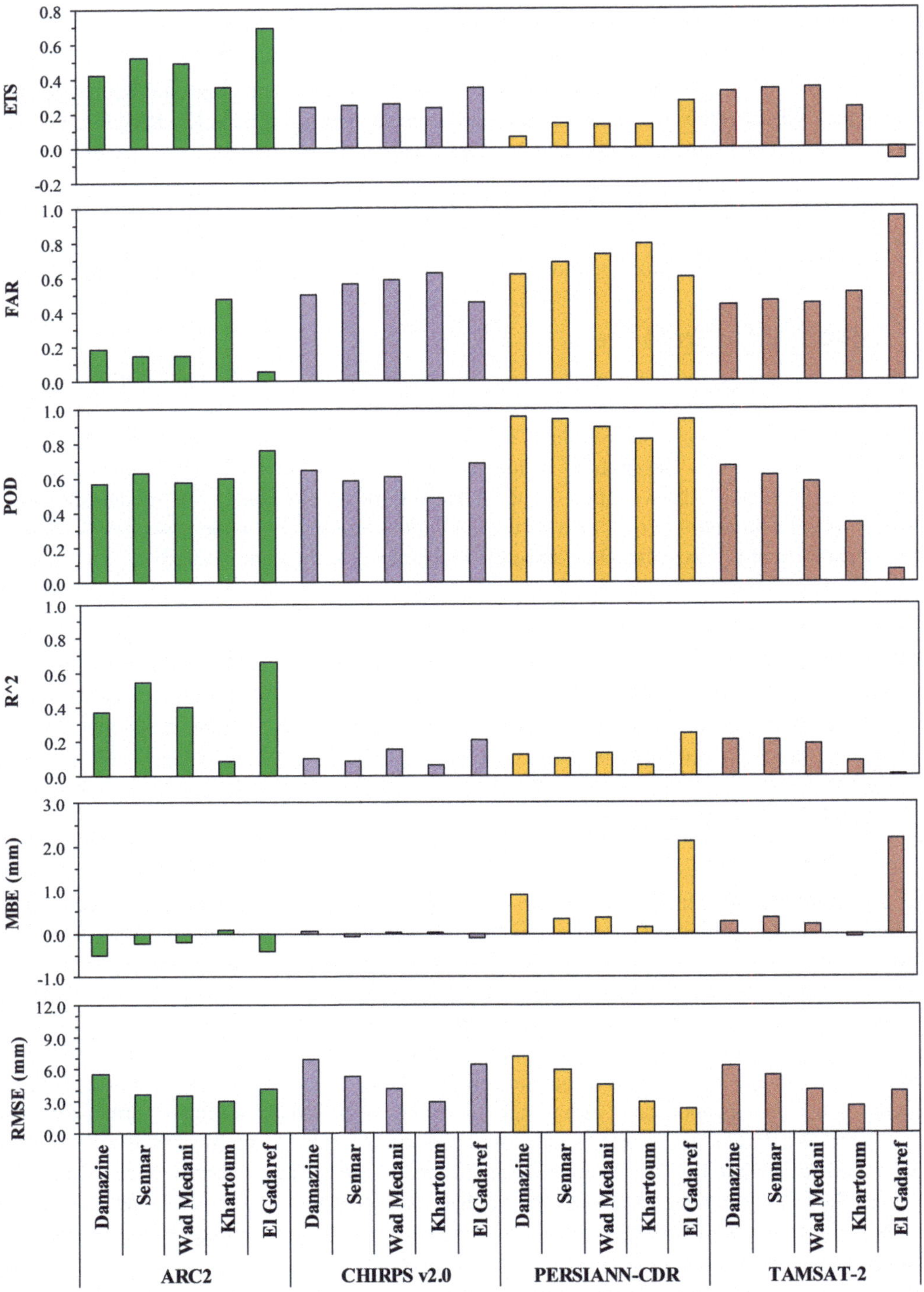

Figure CS7.5 Performance metrics of the evaluated satellite-based rainfall products.

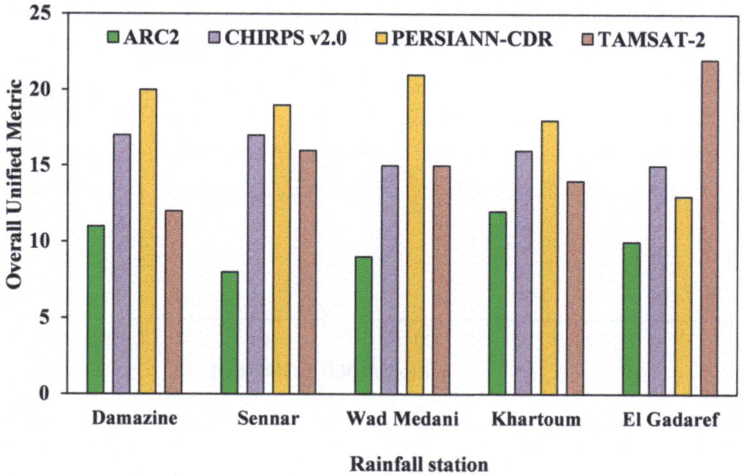

Figure CS7.6 Overall performance of the satellite-based rainfall products in the study area

Roseires and Sennar Dams of around 1.735 BCM, and WEP of around 1575 GWh/BCM. These results indicate that the GERD would further decrease the WEP in Blue Nile Basin Sudan due to an increase in reservoir evaporation at a faster rate than energy generation. Increasing the WEP would require utilizing the untapped hydropower potential from the Roseires and Sennar dams that the GERD would provide as a result of less inter-annual variability in river flow.

The annual irrigation water supply shortage (SS) and the risk of irrigation water supply shortage (RSS) were used in this study as indicators of water–food nexus. The RSS is defined by Wheeler *et al.* (2016) as the percentage of days in the year with SS. Figure CS7.8(b) shows that from 1984 to 2016, the annual irrigation water supply shortage and the annual risk of irrigation water supply shortage ranged from 0 to 0.9 BCM and

Table CS7.1 Model performance at three locations in the study area.

Location	Performance metric	Calibration		Validation	
		Metric value	Ranking	Metric value	Ranking
Roseires Dam	R^2	0.97	Excellent	0.94	Excellent
	NSE	0.96	Excellent	0.98	Excellent
	MPE	0.88	Excellent	0.84	Excellent
Sennar Dam	R^2	0.95	Excellent	0.93	Excellent
	NSE	0.95	Excellent	0.96	Excellent
	MPE	−1.30	Excellent	8.95	Excellent
Khartoum Gauge	R^2	0.90	Excellent	0.91	Excellent
	NSE	0.90	Excellent	0.92	Excellent
	MPE	0.26	Excellent	−13.08	Very good

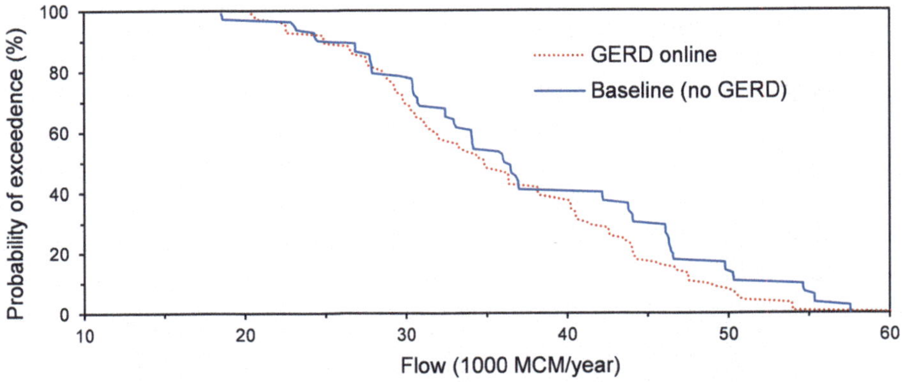

Figure CS7.7 Exceedance probability of the Blue Nile flow at El-diem with and without the GERD.

0 to 27%, respectively. A high association is evident between the annual irrigation water SS, risk of supply shortages (RSS), and river flow (I). The highest shortages and risk occurred in the year 1985 during the drought of the 1980s and 1990s. The blue lines in Figure CS7.8(b) indicate the instances with low irrigation shortages. It can be noticed that the years with above-normal flow conditions often have low irrigation shortages. Moreover, low shortages are evident from the year 2013 due to the heightening of the Roseires Dam. The results show that the steady operation of the GERD would eliminate the irrigation water SS in the study area. This is mainly due to the more regular flow the GERD would provide, as shown in Figure CS7.7.

The annual risk of environmental flow violation (REV) was used to assess the linkages of the ecosystem status with the other nexus components. The REV is the percentage of days in the year where the minimum environmental flow requirements have been violated. Equation (CS7.3) was used to calculate the REV.

$$\mathrm{REV} = \frac{Dv}{DY} \times 100\% \qquad \text{(CS7.3)}$$

where REV is the risk of environmental flow violation in the *i*th year, *Dv* is the number of days where environmental flow requirements have been violated in the *i*th year, and *DY* is the number of days in the *i*th year.

The minimum environmental outflow from the Sennar Dam is 8 MCM/day, according to the Sudanese authorities. The blue-coloured lines in Figure CS7.8(c) mark the instances with a low risk of environmental flow violation in the study area. Figure CS7.8(c) reveals an association between environmental flow provisioning, irrigation water supply, and the hydrologic condition. The highest REV occurred in the year 1985. The heightening of the Roseires Dam reduced the REV significantly. The analysis of the steady-state operation of the GERD showed that the dam would eliminate any risk of environmental flow violation. This is due to the more regular flow that the GERD would provide.

CS7.5 CONCLUSIONS AND WAY FORWARD

The integrative approach of the WEFE Nexus offers an opportunity to utilize resources more efficiently by taking into account their interconnections in management and planning. However, there remains an operationalization gap due to the lack of tools and assessment metrics (Liu *et al.*, 2017). This study is an

(a)

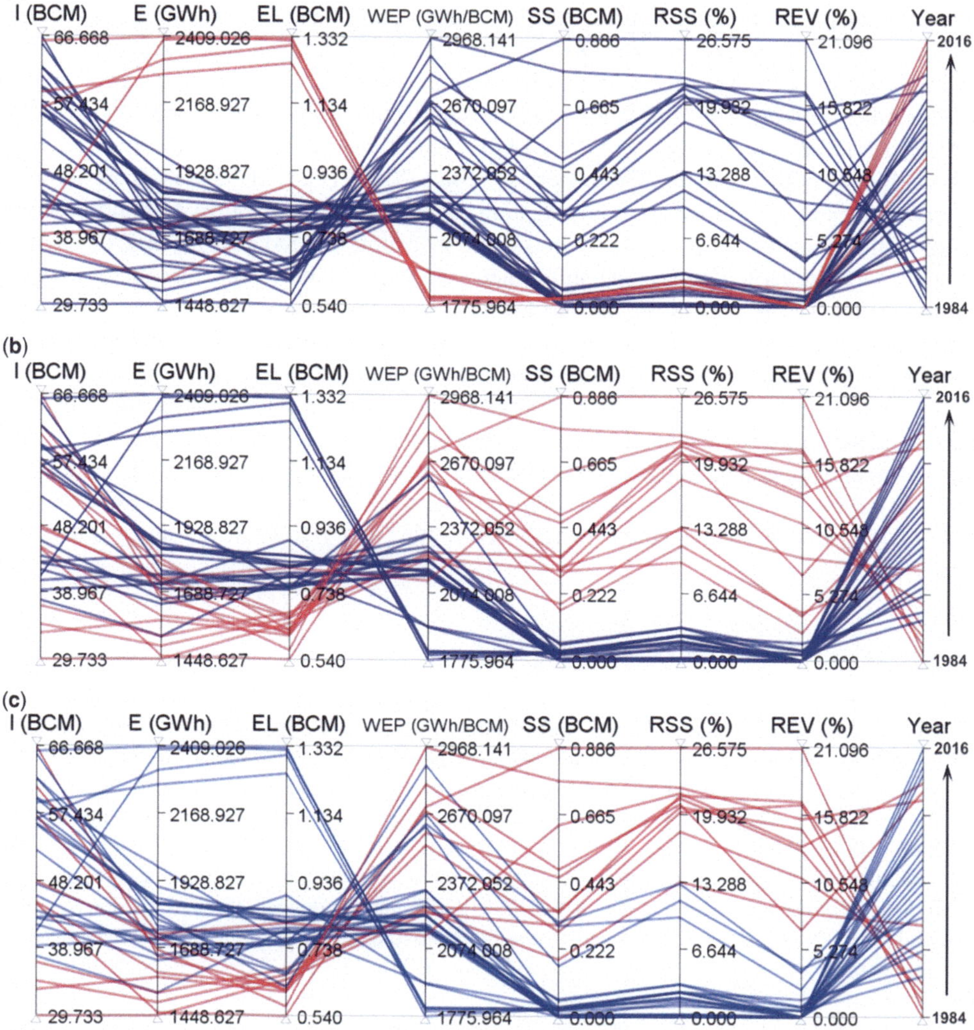

Figure CS7.8 Parallel plots of the historic annual (1984–2016) WEFE Nexus indicators: (a) blue- and red-coloured lines represent years with high and low WEP, respectively; (b) blue- and red-coloured lines represent years with low and high irrigation water shortages, respectively; (c) blue- and red-coloured lines represent years with low and high environmental flow violation, respectively. I = inflow; E = energy generation; EL = evaporation losses; WEP = water–energy productivity; SS = irrigation supply shortages; RSS = risk of irrigation supply shortage; REV = risk to environmental flow violation.

attempt to quantify the interlinkages of the water, energy, food, and ecosystem in a resource stressed and data-scarce region. The study region is the Blue Nile Basin downstream in Sudan. This region has experienced past persistent hydrological droughts, recent dam development, and will soon experience a major dam development in the upstream.

A daily rainfall–runoff and water allocation model was developed for the study region to quantify some WEFE Nexus indicators. The model covers the 1984–2016 period and includes the major water-related infrastructures and their operating rules. Due to data scarcity in the study region, four satellite-based rainfall products have been evaluated using a pixel-to-point approach, and the best performing one was used as a boundary condition to model the rainfall–runoff component of the model. The results show that the African Rainfall Climatology Version 2 (ARC2) has the best performance compared to the other evaluated satellite rainfall products. The historical (1984–2016) nexus indicators show an association between environmental flow provisioning, irrigation water supply, and the hydrologic condition. The heightening of the Roseires Dam in 2013 reduced the irrigation SS, reduced the risk of environmental flow violation, increase hydropower generation, increased evaporation losses, and increased the water–energy productivity. The results show that the Grand Ethiopian Renaissance Dam would eliminate the risk of environmental flow violation, eliminate the irrigation SS, increase hydropower generation, increase evaporation losses, and reduce the inter-annual variability in river flow.

The results of this study show a promising role that satellite data can play in data-scarce regions. A more extensive evaluation of all the available satellite-based rainfall products would be needed to exploit the potential of this emerging data source fully. The analysis conducted herein uses the index-sequential method to analyse future scenarios. However, this method does not take into account non-stationarity in the climate system. Future studies could focus on examining the WEFE Nexus under transient climate conditions.

REFERENCES

Al-Saidi M. and Elagib N. (2017). Towards understanding the integrative approach of the water, energy and food nexus. *Science of the Total Environment*, **574**, 1131–1139. https://doi.org/10.1016/j.scitotenv.2016.09.046.

Ashouri H., Hsu K., Sorooshian S., Braithwaite D., Knapp K., Cecil D., Nelson B. and Prat O. (2015). PERSIANN-CDR: daily precipitation climate data record from multisatellite observations for hydrological and climate studies. *Bulletin of the American Meteorological Society*, **96**, 69–83. https://doi.org/10.1175/BAMS-D-13-00068.1.

Basheer M. and Elagib N. A. (2018a). Sensitivity of water-energy nexus to dam operation: A water-energy productivity concept. *Science of the Total Environment*, **616–617**, 918–926. https://doi.org/10.1016/j.scitotenv.2017.10.228.

Basheer M. and Elagib N. A. (2018b). Performance of satellite-based and GPCC 7.0 rainfall products in an extremely data-scarce country in the Nile basin. *Atmospheric Research*, **215**, 128–140. https://doi.org/10.1016/j.atmosres.2018.08.028.

Basheer M., Wheeler K. G., Ribbe L., Majdalawi M., Abdo G. and Zagona E. A. (2018). Quantifying and evaluating the impacts of cooperation in transboundary river basins on the water-energy-food nexus: the Blue Nile Basin. *Science of the Total Environment*, **630**, 1309–1323. https://doi.org/10.1016/j.scitotenv.2018.02.249.

Bazilian M., Rogner H., Howells M., Hermann S., Arent D., Gielen D., Komor P., Steduto P., Mueller A., Tol R. and Yumkella K. (2011). Considering the energy, water and food nexus: towards an integrated modelling approach. *Energy Policy*, **39**, 7896–7906. https://doi.org/10.1016/j.enpol.2011.09.039.

Bhattacharyya S., Bugatti N. and Bauer H. (2015). A bottom-up approach to the nexus of energy, food and water security in the Economic Community of West African States (ECOWAS) region.

Chai T. and Draxler R. (2014). Root mean square error (RMSE) or mean absolute error (MAE)? Arguments against avoiding RMSE in the literature. *Geoscientific Model Development*, **7**, 1247–1250. https://doi.org/10.5194/gmd-7-1247-2014.

Diem J., Hartter J., Ryan S. and Palace M. (2014). Validation of satellite rainfall products for western Uganda. *Journal of Hydrometeorology*, **15**, 2030–2038. https://doi.org/10.1175/JHM-D-13-0193.1

DIU (2016). Roseires heightening project [WWW Document]. Ministry of Water Resources and Electricity of Sudan. URL http://www.roseiresdam.gov.sd/en/raising_project.htm (accessed 27 January 2019).

Ebert E., Janowiak J. and Kidd C. (2007). Comparison of near-real-time precipitation estimates from satellite observations and numerical models. *Bulletin of the American Meteorological Society*, **88**, 47–64. https://doi.org/10.1175/BAMS-88-1-47.

Elagib N. A., Gayoum Saad S. A., Basheer M., Rahma A. E. and Gore E. D. L. (2019). Exploring the urban water-energy-food nexus under environmental hazards within the Nile. *Stochastic Environmental Research and Risk Assessment*. https://doi.org/10.1007/s00477-019-01706-x.

Endo A., Tsurita I., Burnett K. and Orencio P. M. (2017). A review of the current state of research on the water, energy, and food nexus. *Journal of Hydrology: Regional Studies*, **11**, 20–30. https://doi.org/10.1016/j.ejrh.2015.11.010.

Funk C., Peterson P., Landsfeld M., Pedreros D., Verdin J., Rowland J., Romero B., Husak G., Michaelsen J. and Verdin A. (2014). A quasi-global precipitation time series for drought monitoring. *U.S. Geological Survey Data Series*, **832**, 4. https://doi.org/110.3133/ds832.

HEC (2008). Hydrologic Modeling System HEC-HMS Applications Guide. The US Army Corps of Engineers - Hydrologic Engineering Center, Washington, D.C.

Hoff H. (2011). Understanding the Nexus. Background Paper for the Bonn 2011 Conference: The Water, Energy and Food Security Nexus. Stockholm Environment Institute, Stockholm.

Howells M., Hermann S., Welsch M., Bazilian M., Segerström R., Alfstad T., Gielen D., Rogner H., Fischer G., van Velthuizen H., Wiberg D., Young C., Roehrl R., Mueller A., Steduto P. and Ramma I. (2013). Integrated analysis of climate change, land-use, energy and water strategies. *Nature Climate Change*, **3**, 621–626. https://doi.org/10.1038/nclimate1789.

IWMI (2012). The Nile River Basin water, Agriculture, Governance and Livelihoods, 1st ed. Routledge, Oxon.

Kendall D. and Dracup J. (1991). A comparison of index-sequential and AR(1) generated hydrologic sequences. *Journal of Hydrology*, **122**, 335–352. https://doi.org/10.1016/0022-1694(91)90187-M.

Keskinen M., Someth P., Salmivaara A. and Kummu M. (2015). Water-energy-food nexus in a transboundary river basin: the case of Tonle Sap Lake, Mekong River Basin. *Water*, **7**, 5416–5436. https://doi.org/10.3390/w7105416.

Kibaroglu A. and Gürsoy S. (2015). Water–energy–food nexus in a transboundary context: the Euphrates–Tigris river basin as a case study. *Water International*, **40**, 824–838. https://doi.org/10.1080/02508060.2015.1078577.

Legates D. and McCabe G., Jr (1999). Evaluating the use of "goodness of fit" measures in hydrologic and hydroclimatic model validation. *Water Resources Research*, **35**, 233–241. https://doi.org/10.1029/1998WR900018.

Liu J., Yang H., Cudennec C., Gain A. K., Hoff H., Lawford R., Qi J., de Strasser L., Yillia P. T. and Zheng C. (2017). Challenges in operationalizing the water–energy–food nexus. *Hydrological Sciences Journal*, **62**, 1714–1720. https://doi.org/10.1080/02626667.2017.1353695.

Maidment R., Grimes D., Allan R., Tarnavsky E., Stringer M., Hewison T., Roebeling R. and Black E. (2014). The 30 year TAMSAT African Rainfall Climatology And Time series (TARCAT) data set. *Journal of Geophysical Research: Atmospheres*, **119**, 10619–10644. https://doi.org/10.1002/2014JD021927.

MoIHES (1977). Blue Nile Waters Study Phase IA: Availability and use of Blue Nile Water. Khartoum.

MoIHPS (1966). Roseires Dam. Khartoum.

Novella N. and Thiaw W. (2013). African rainfall climatology version 2 for famine early warning systems. *Journal of Applied Meteorology and Climatology*, **52**, 588–606. https://doi.org/10.1175/JAMC-D-11-0238.1.

Ouarda T., Labadie J. and Fontane D. (1997). Indexed sequential hydrologic modeling for hydropower capacity estimation. *Journal of the American Water Resources Association*, **33**, 1337–1349.

Pittock J., Dumaresq D. and Bassi A. (2016). Modeling the hydropower–food nexus in large river basins: A Mekong case study. *Water*, **8**. https://doi.org/10.3390/w8100425.

R Core Team (2015). R: A Language and Environment for Statistical Computing [WWW Document]. URL at: https://www.R-project.org/ (accessed 27 June 2017).

Roseires Dam Heightening Unit (2005). Updating economic and financial viability study of dam heightening. Khartoum.

Stamou A. and Rutschmann P. (2018). Pareto optimization of water resources using the nexus approach. *Water Resources Management*. https://doi.org/10.1007/s11269-018-2127-x.

Stern M., Flint L., Minear J., Flint A. and Wright S. (2016). Characterizing changes in streamflow and sediment supply in the Sacramento River Basin, California, using hydrological simulation program—FORTRAN (HSPF). *Water*, **432**, 1–21. https://doi.org/10.3390/w8100432.

Strasser L., Lipponen A., Howells M., Stec S. and Bréthaut C. (2016). A methodology to assess the water energy food ecosystems nexus in transboundary river basins. *Water*, **8**. https://doi.org/10.3390/w8020059.

Toté C., Patricio D., Boogaard H., van der Wijngaart R., Tarnavsky E. and Funk C. (2015). Evaluation of satellite rainfall estimates for drought and flood monitoring in Mozambique. *Remote Sensing*, **7**, 1758–1776. https://doi.org/10.3390/rs70201758.

UN (2015a). World Population Prospects: Key Findings and Advance Tables (No. ESA/P/WP.241). New York.

UN (2015b). The Millennium Development Goals Report, United Nations. New York.

UNDP (2015). Human Development Report 2015: Work for Human Development. New York.

Wheeler K. G., Basheer M., Mekonnen Z., Eltoum S., Mersha A., Abdo G., Zagona E., Hall J. and Dadson S. (2016). Cooperative filling approaches for the Grand Ethiopian Renaissance Dam. *Water International*, **8060**, 1–24. https://doi.org/10.1080/02508060.2016.1177698.

Wheeler K., Hall J., Abdo G., Dadson S., Kasprzyk J., Smith R. and Zagona E. (2018). Exploring cooperative transboundary river management strategies for the Eastern Nile Basin. *Water Resources Research*, 9224–9254. https://doi.org/10.1029/2017WR022149.

Zagona E., Fulp T., Shane R., Magee T. and Goranflo H. M. (2001). Riverware: a generalized tool for complex reservoir system modeling. *Journal of the American Water Resources Association*, **37**, 913–929.

Towards the circularization of the energy cycle by implementation of hydroelectricity production in existing hydraulic systems

M. Marence, S. Lemessa Tesgera and M. J. Franca

River Basin Development, IHE Delft Institute for Water Education, Delft, The Netherlands

ABSTRACT

Hydropower potential is widely recognizable as the most important low-carbon electricity source. Nevertheless, implementation struggles with restrictions and negative image. High construction cost and social and environmental impact additionally discourage new developers. Installation of new hydropower plants in existing water infrastructures, without hindering their primary functions, is an attractive solution, saving construction costs and minimizing environmental impact. Moreover, it may contribute to the reduction of the global footprint of existing hydraulic infrastructures. Here, we discuss several possible solutions to implement hydropower production systems in existing hydraulic infrastructures, referring to technic and economic feasibility and how these open avenues to global footprint reduction. A case study of the implementation of a hydropower plant associated with an existing irrigation system in Ethiopia is presented. In this case the electrical energy can be extracted from an existing system without hindering its main function linked to food production, representing an added value for the owner, for the local population, and for the environment. The further development and investigation of synergetic approaches as the ones here presented are required to achieve the sustainable development goals, according to which water availability must be secured for both food and low-carbon energy production.

Keywords: hydroelectricity, energy cycle, Awash river, Tandaho, Ethiopia

CS8.1 INTRODUCTION

The link between water, energy, and food is inextricable since water is essential for human direct consumption, crops production, and hydropower electricity provision. Population growth, rapid urbanization, changing diets, economic development, and climate change impose a rising demand and stress for food and energy production (WWAP: World Water Assessment Programme, 2012, 2014). Thus, synergetic approaches to respond to both food and energy needs, requiring an integrated management of water, are essential to respond to the WEF Nexus, and in particular to attain the SDGs 2, 6, and 7.

Hydropower is one of the oldest renewable energy sources and by far the most important source of low-carbon energy in the world. Hydropower relies on water passing through turbines to generate electricity by conversion of mechanical energy. Inserted in the global energy production mix, hydropower also offers the opportunity to serve as a major source of global energy storage (Barbour *et al.*, 2016). Installation of hydropower plants in the greenfield imposes environmental and social impacts which have to be tackled before construction to achieve a sustainable and impact friendly energy source.

In human infrastructures, the hydraulic energy contained in flowing water is frequently purposely dissipated for operational reasons; that is, the energy is simply lost or wasted. The loss of mass (water leakage) and the excess of momentum (which may provoke pipe bursts for instance) are well perceived by humans. However, the waste of energy in hydraulic systems has no visual or sensorial expression, which justifies the common qualification given to this availability of energy, the 'hidden treasure'. Hence, one of the main directions for the decarbonization of the sources of energy, in the context of hydroelectricity, must be the fight against this energy waste which is hidden in the existing hydraulic infrastructure.

The development of alternative hydropower plants incorporated in existing hydro-technical structures, used primarily for other purposes, represents an additional and profitable solution for energy generation with low impact (Marence *et al.*, 2016). Furthermore, the use (or re-use) of the so-called wasted energy contributes to the (partial) circularization of the energy cycle in hydraulic infrastructures contributing to reduce (or even revert in a long term) the anthropogenic impact of such systems.

Hydropower incorporated in the existing hydraulic infrastructures, where electricity generation was not a primary objective, is an attractive solution for implementation in developed and developing energy systems from different perspectives:

- Environmental and social boundaries on the existing and operating hydro-technical structures are known, defined, and mostly accepted.
- Installation of the new hydropower in existing system save construction costs and large infrastructural works, also reducing construction time and greenhouse gas (GHGs) emissions during the construction phase.
- Lower construction costs with shorter payback ratio make the solutions attractive for investment.
- Reduced uncertainty to the resource availability and known boundary conditions in terms of obtainable water flow and energy head.
- Reduction of the operational costs by synergies in the operation.
- Modifications for the additional purpose gives the possibility for social and environmental improvements.

Identification and implementation of hydropower in such non-traditional options can be seen as the search for the so-mentioned hidden treasure. Possible solutions may be found in: municipal and agricultural water systems (respectively in urban and rural environments); existing dams and hydropower plants; and industrial processes and hydraulic circulation systems. Some of these options are shown in Figure CS8.1.

CS8.2 KEY CONCEPTS FOR ADDITIONAL HYDROPOWER IMPLEMENTATION

CS8.2.1 Municipal and agricultural water systems

Urban water supply systems are complex large technological systems (Hornberger & Hess, 2015), the purpose of which is to extract, treat, convey, and distribute water to consumers. As such, they are

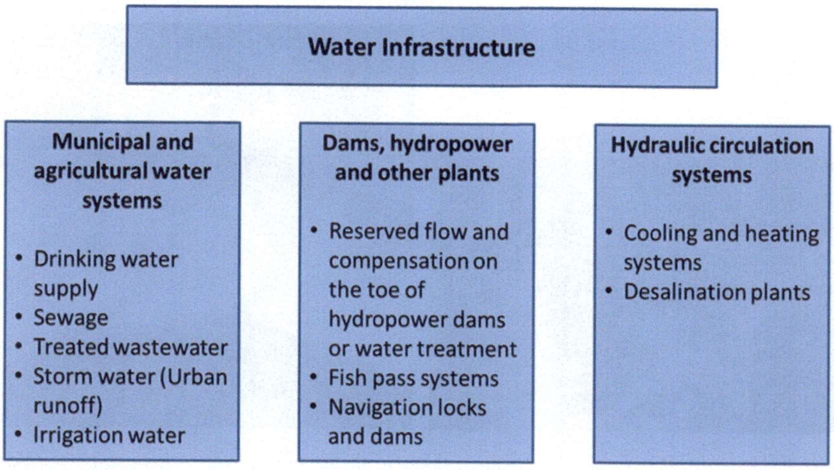

Figure CS8.1 Water infrastructures suitable for additional hydropower implementation (Marence *et al.*, 2016).

energy-consuming systems with their own carbon and energy footprint. The search for low-carbon cities usually neglects the role and the potential for electricity production which is (often literally) buried in water supply networks, a built and essential element of the urban settlements. The potential is stronger in mountainous or hilly regions, where water intakes are situated at higher altitude and excess pressure must be reduced before consumption. This reduction is performed by energy dissipation in specially constructed valves saving system pipes from bursting and permitting the operational use of water.

The pressure reduction by the installation of turbines in drinking water supply systems is already used, for instance, in Alpine areas and in steep mountainous islands (Vieira & Ramos, 2008). The installation of turbines in conveyance pipelines is technically easier compared with a distribution network because flow and pressure in the first are less fluctuating. Drinking water turbines must satisfy sanitary requirements without any influence on the water quality and allowing full water supply independently of the turbine operation.

Bölli and Feibe (2015) refer to an overlying potential of hydropower production of about 35 GWh per year in the water supply system of SONEDE, the Tunisian water supply utility. The study by Samora *et al.* (2016), which focused on the hydropower potential of the water supply network of the city of Fribourg, in Switzerland, shows that about 40% of the energy spent for pumping water for this system is wasted by the network. According to Bergkamp (2015), the energy to supply treated water and to dispose wastewater is responsible for 3–8% of the global GHGs emissions. Taking into account that urban settlements produce about 70% of the carbon emissions in the world (Wade, 2014), one may argue that there exists a large potential for reducing global GHGs emission with such adaptations in water supply conveyance pipelines and distribution networks.

Also the municipal sewage systems have potential for hydropower production. In this case, the natural head that generally exists between the residential area and the disposal of treated water in nature can be used. In the case of untreated water, the energy system may need to deal with solid debris. Recently, Bousquet *et al.* (2017) estimated the hydropower potential of the use of wastewater systems in Switzerland and identified 19 profitable sites with a total 9.3 GWh/year of potential electrical energy production.

(a) (b)

Figure CS8.2 Example of pressure reducers installed at the dam toe, upstream irrigation networks: (a) Odivelas dam; and, (b) Gostei dam, both in Portugal. *Photo*: © Mario Franca.

Irrigation systems are mostly built with a function to maximize the irrigated area and transport water as far as possible. In the toe of irrigation dams, at the upstream section of the irrigation network, the excess of energy needs often to be dissipated by special valves (Figure CS8.2). In some cases, where the downstream channel network presents drops, the difference of level in these singularities may be used for energy extraction. A case study is presented in Section 3 regarding an irrigation system in Ethiopia, where examples of these two types of energy extraction were analysed.

CS8.2.2 Dams, hydropower, and other plants

Dams store water for different uses and requirements. More than 70% of world dams are built for a single purpose and irrigation is the most common use. Just 17.4% of world dams are built for hydropower and energy generation (ICOLD, 2017). Most of the dams built for other purposes have the possibility for additional energy generation. The potential and appeal of such solutions is demonstrated by the case of USA where more than 80,000 non-powered dams have been detected with a total potential of additional 12 GW (Hadjerioua *et al.*, 2012).

Existing hydropower plants have potential for additional upgrade that could be done by refurbishment and installation of modern equipment with higher efficiency and higher load factor. Also, the implementation of energy converters associated with ecological measures, such as fish ladders and bypasses, and ecological minimal flow releases, is desirable since in these energy is often available and needs to be dissipated.

Additional possibilities are the installation of hydropower plants in ship navigation locks where the flowing water needed for filling and emptying these may be used for energy generation. Weirs controlling navigation levels, are another possible source for energy generation. On the river Waal, in the Netherlands, three navigation weirs include hydropower plants (Manders *et al.*, 2016).

CS8.2.3 Hydraulic circulation systems

Similar to drinking water systems, industrial cooling or heating systems and water processing systems can result in a pressure excess that can be recovered through hydro turbines instead of being lost in energy dissipators.

The water abstraction for the cooling systems of thermal power plants represents from 10 to 50% (depending on the country and region) of the total freshwater withdrawals (WWAP: World Water Assessment Programme, 2014). This water, after temperature exchange, is released back into the river. The head difference between the river and the process station depends on the configuration and river flood characteristics, but in most case stays unexplored.

Some desalination plants use reverse osmosis to separate water from dissolved salts through semipermeable membranes under high pressure (from 40 to 80 bars). The residue of water containing salt, still at high pressure, could be flown through a turbine in order to recover part of the energy used for the initial compression.

CS8.3 CASE STUDY: IMPLEMENTATION OF HYDROPOWER PRODUCTION IN THE EXISTING IRRIGATION SYSTEMS OF THE KESSEM AND TANDAHO RESERVOIRS, IN ETHIOPIA

Often in developing countries, the installed capacity for hydroelectricity generation falls far short from their potential. In particular, Ethiopia has a capacity of 1500 MW for small hydropower with 6 MW installed capacity, thus only 0.4% of its potential (United Nations Industrial Development Organization & International Center on Small Hydro Power, 2016). According to the World Bank, in this country only 22.2% of the population in rural areas has access to electricity. Without provision of electricity to this large population, sustainable development and eradication of poverty and illiteracy is hindered. This situation urges for opportunities for electrical energy production, namely in existing water infrastructures such as reservoirs and irrigation canals which are near the communities.

The Kessem and Tandaho irrigation reservoirs are located in the lower Awash river basin in Ethiopia (Figure CS8.3). The reservoirs collect and store rainfall and river flow water during the rainy season and a controlled outlet flow is used for irrigation of agricultural land, used mostly for sugar cane production. Estimations of the potential for electrical energy production were made for two types of hydropower plants corresponding to: the use of the head available at the toe of the Kessem, before this is released to the irrigation network; the use of existing drops in the 44 km irrigation channel downstream of the Tandaho dam.

The Kessem dam has a height of about 90 m. The potential head and flow for hydropower production downstream Kessem dam were estimated from a simulation of a mean year of operation of the irrigation system. The storage water routing is performed as a function of the depth–volume curve of the reservoir, taking into account monthly irrigation needs and inflowing hydrological data from more than 30 years. A possible yearly energy production of 31.8 GWh was estimated.

The Tandaho canal (Figure CS8.4) is regulated based on crop water requirements. The irrigation network downstream of the Tandaho dam, after satisfying minimal flow regulations in the river, is diverted to a concrete artificial channel with a minimum operating discharge of 18 m^3/s. The channel has a continuous slope of 0.01% with several concentrated drops of 1.2–3.0 m. Six of these hydraulic drops may be used to extract energy by means of a screw turbine, with a reference production value per drop of 1.75 GWh per year. In the study area two small villages and sugar factories are near to the irrigation structures.

Both solutions associated with the two irrigation systems belonging to the lower Awash valley could produce as much as 42 GWh of hydroelectricity, which could supply the national grid or be used locally to support regional development or to support agricultural production. Using the data in Bruckner *et al.* (2014) as reference as well as the study by Zhang *et al.* (2015), where CO_2 emissions of small

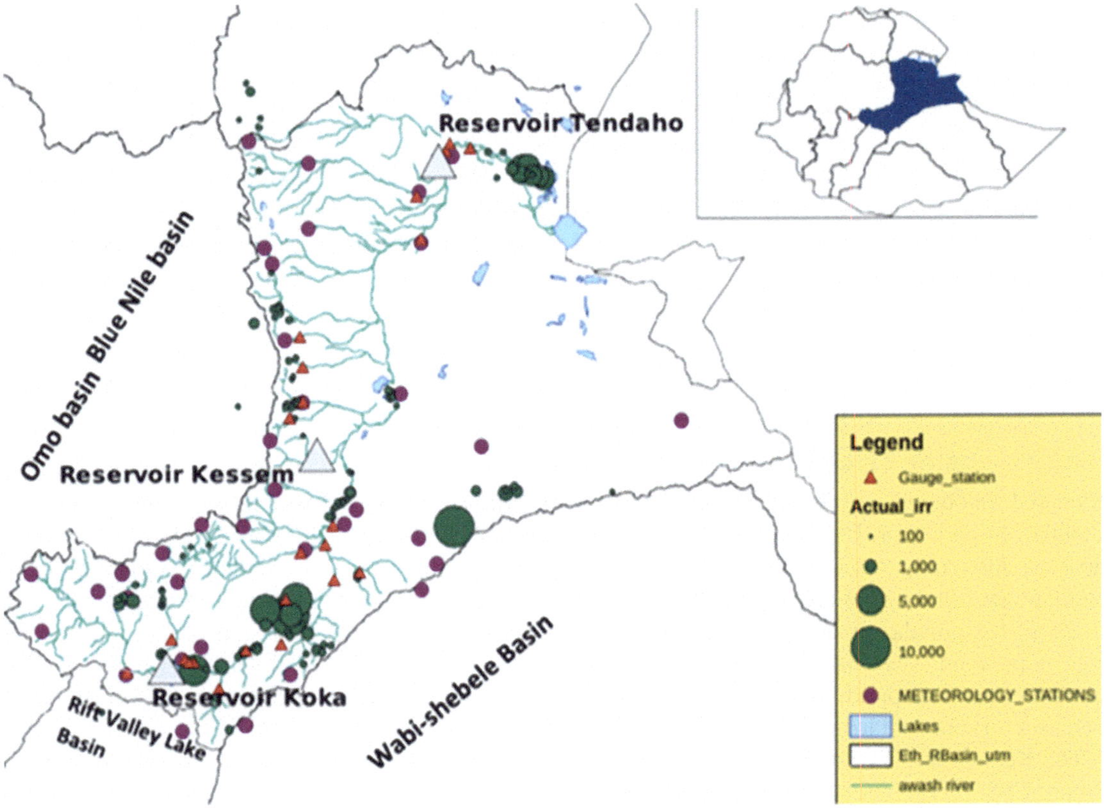

Figure CS8.3 Schematic layout of Awash River basin gauging stations, irrigation capacity and meteorological stations (Müller *et al.*, 2016).

Figure CS8.4 Head regulator of Tandaho irrigation canal and water diversion to the command area. *Photo*: © Mario Franca.

hydropower plants in southwest China were comprehensively discussed, a total reduction of more than 30,000 ton of CO_2 emitted per year could be expected with the implementation of such hydraulic generators when compared to fossil-fuel sources. In both cases, CO_2 emissions associated with the construction of the plant and to the presence of the reservoir are minimized. The installation of the hydropower plant associated with the irrigation systems would thus contribute to reduce their global impact.

The implementation of energy extraction plants in existing hydraulic structures must consider the uninterrupted operation of the facility and the minimization of the original function of the irrigation system which is agricultural production. The original function and the crops needs dictate the operation and available water for energy production, and priority for the use of the available water stored in the reservoir must be given to irrigation.

CS8.4 DISCUSSION, CONCLUSIONS, AND RECOMMENDATIONS

Hydraulic systems supporting food production, water and sanitation supply, industry and hydroelectricity production have common engineering mechanisms of storage and conduction of water. Thus, the implementation of synergetic multi-purpose systems infolding the several uses here mentioned is the best way to respond to the water–food–energy Nexus, at the same time contribution to the fulfillment of the SDGs, in particular SDGs 2, 6, and 7.

Hydropower rehabilitation and optimization, together with the installation of power plants in existing hydraulic structures (the so-called hidden treasure), need to be tapped. There is a considerable potential for rehabilitation, life extension, upgrading and optimization of existing hydropower facilities by improving energy efficiency and at the same time guaranteeing the safety of ageing plants. Rehabilitation and upgrading of existing plants can be made by using modern technologies and comprehensive planning while minimizing environmental and social impacts.

Moreover, and as shown here with examples in urban and rural areas, large infrastructures and industrial applications, the installation of hydroelectricity production equipment in existing hydraulic systems gives additional possibilities for low-carbon energy generation. The presented example shows a high potential for additional energy production by hydropower installed in an irrigation infrastructure. The existing structures may often be used with minimal or no extra social and environmental impacts. Furthermore, these solutions can contribute to reduce or even cancel the negative impact of the existing infrastructure as shown above.

Investigation of synergetic approaches as the ones discussed here are required to achieve the SDGs, according to which water availability must be secured for both food and low-carbon energy production. Research on the optimization and management of multi-purpose systems as well as on technological solutions aiming at the (partial) circularization of the energy cycle in hydraulic systems, is urgently required.

REFERENCES

Barbour E., Wilson I. A. G., Radcliffe J., Ding Y. and Li Y. (2016). A review of pumped hydro energy storage development in significant international electricity markets. *Renewable and Sustainable Energy Reviews*, **61**, 421–432. https://doi.org/10.1016/j.rser.2016.04.019 (last accessed July 2021).

Bergkamp G. (2015, April). Energy and carbon neutral water cities. IWA Blog. Retrieved from http://www.iwa-network.org/energy-and-carbon-neutral-water-cities/ (last accessed July 2021).

Bölli M. and Feibe H. (2015). Low hanging fruits: hydropower in existing water infrastructure. *ESI Africa*, **1**, 76–77.

Bruckner T., Bashmakov I. A., Mulugetta Y., Chum H., de la Vega Navarro A., Edmonds J., Hertwich E., Honnery D., Infield M., Kainuma S., Khennas S., Kim H. B., Nimir K., Riahi N., Strachan R. and Wiser X. Z. (2014). Climate

Change 2014: Mitigation of climate change. IPCC Fifth Assessment Report, 527–532. https://doi.org/10.1017/CBO9781107415416 (last accessed July 2021).

Bousquet C., Samora I., Manso P., Rossi L., Heller P. and Schleiss A. J. (2017). Assessment of hydropower potential in wastewater systems and application to Switzerland. *Renewable Energy*, **113**, 64–73.

Hadjerioua B., Wei Y. and Kao S.-C. (2012). An assessment of energy potential at non-powered dams in the United States. U.S. Department of Energy. https://doi.org/10.2172/1039957 (last accessed July 2021).

Hornberger G. and Hess D. (2015). Water conservation and hydrological transitions in cities in the United States. *Water Resources*, **51**(6), 4635–4649. Retrieved from http://onlinelibrary.wiley.com/doi/10.1002/2015WR016943/full (last accessed July 2021).

ICOLD (2017). Dams & the world's water. Paris: International Commission on Large Dams. Retrieved from www.icold-cigb.org (last accessed July 2021).

Manders T. N., Höffken J. I. and Van Der Vleuten E. B. A. (2016). Small-scale hydropower in the Netherlands: problems and strategies of system builders. *Renewable and Sustainable Energy Reviews*, **59**, 1493–1503. https://doi.org/10.1016/j.rser.2015.12.100 (last accessed July 2021).

Marence M., Ingabire J. S. and Taks B. (2016). Integration of hydropower plant within an existing weir – 'a hidden treasure'. In: Sustainable hydraulics in the era of global change, M. Erpicum, S. Dewals, B. Archambeau and P. Pirotton (eds.), Taylor & Francis Group, London, pp. 974–978.

Müller R., Gebretsadik H. Y. and Schütze N. (2016). Towards an optimal integrated reservoir system management for the Awash River Basin, Ethiopia. In: 7th International Water Resources Management Conference of ICWRS. Bochum, p. IWRM2016-95-2.

Samora I., Manso P., Franca M., Schleiss A. and Ramos H. (2016). Energy recovery using micro-hydropower technology in water supply systems: the case study of the city of Fribourg. *Water*, **8**(8), 344. https://doi.org/10.3390/w8080344 (last accessed July 2021).

United Nations Industrial Development Organization, and International Center on Small Hydro Power (2016). The World Small Hydropower Development Report. Retrieved from http://www.unido.org/fileadmin/user_media_upgrade/What_we_do/Topics/Energy_access/WSHPDR_2013_Executive_Summary.pdf (last accessed July 2021).

Vieira F. and Ramos H. (2008). Hybrid solution and pump-storage optimization in water supply system efficiency: a case study. *Energy Policy*, **36**(11), 4142–4148.

Wade L. (2014). Giving cities a road map to reducing their carbon footprint. Retrieved from http://www.sciencemag.org/news/2014/12/giving-cities-road-map-reducing-their-carbon-footprint (last accessed July 2021).

WWAP: World Water Assessment Programme (2012). The United Nations World Water Development Report 4: Managing Water under Uncertainty and Risk. (UNESCO, Ed.), UN Water Reports. Paris. Retrieved from http://unesdoc.unesco.org/images/0021/002156/215644e.pdf (last accessed July 2021).

WWAP: World Water Assessment Programme (2014). The United Nations World Water Development Report 2014: Water and Energy (UNESCO, Vol. 1). PAris. https://doi.org/978-92-3-104259-1 (last accessed July 2021).

Zhang J., Luo C.-Y., Curtis Z., Deng S., Wu Y. and Li Y. (2015). Carbon dioxide emission accounting for small hydropower plants – A case study in southwest China. *Renewable and Sustainable Energy Reviews*, **47**, 755–761. https://doi.org/10.1016/j.rser.2015.03.027 (last accessed July 2021).

Status of geothermal industry in east African countries

A. Battistelli[1], E. Crestaz[2] and C. Carmona-Moreno[2]

[1]Independent Reservoir Engineer, Via Aldo Moro, 4 I-60033 Chiaravalle (AN), Italy
[2]DG Joint Research Centre, Via E. Fermi, T.P. 440, I-21027 Ispra (VA), Italy

ABSTRACT

The role and potential of geothermal energy in 11 countries crossed by the East African rift system, geographically extending from Eastern to Southern Africa, has been reviewed. The focus is on geothermal resources aimed at generating electric power by using either flashing or Organic Rankine Cycle plants with geothermal fluids extracted from medium to high temperature hydrothermal systems. The business models implemented are discussed, in relation with the peculiar features of the geothermal energy which is characterized by important initial investments and limited operating and maintenance expenditures, as most of the renewable energy sources, but having peculiar remarkable mining risks mainly related to the exploration drilling phase. Constraints delaying a more widespread use of geothermal energy for electric power generation in East Africa are analysed, together with the role of international and financial institutions in providing funds and risk mitigation opportunities, support in capacity building and the development of national legal frameworks needed for an improved and faster development of geothermal resources in East Africa. A review of the present status of geothermal development initiatives underway in each of the 11 countries is presented, looking at the possible role of geothermal resources within the electric energy market in each country.

Keywords: geothermal energy, Geothermal potential, East Africa, East African Rift System

CS9.1 INTRODUCTION

In the light of the debate on energy production and on renewable energy in Africa, a decision was taken to integrate the ACE WATER2 overall framework by investigating the role and potential of geothermal energy, that looks like promising in several countries along the East African Rift System (EARS), geographically extending from Eastern to Southern Africa. The general objective of the study was to present the state of the art on the geothermal resource development in East African Countries (Eritrea, Djibouti, Ethiopia, Kenya, Uganda, Rwanda, Burundi, Tanzania, and Comoros) and in two Southern

African countries (Malawi and Zambia), for the sake of simplicity all collectively referred to as 'East African countries'.

The focus of the study is on geothermal activities aimed at generating electric power by using either flashing or organic Rankine cycle (ORC) plants with geothermal fluids extracted from medium-to-high temperature hydrothermal systems.

Geothermal energy is a renewable energy characterized by: low environmental impact and GHGs emissions when compared to energy generated using fossil fuels; quite constant generation output independent from weather conditions, which makes it particularly suitable for base load electric generation; high initial capital costs and low operating and management expenditures; remarkable mining risks mainly related to the results of exploratory drilling phase.

Geothermal resources, found in the core of the Earth crust, are presently exploited for both electric power generation and for direct uses. Favourable geodynamic environments allow founding exploitable geothermal systems at economic and technical feasible depths. Apart for the utilization of low-temperature resources (<100°C) only made for direct uses, the generation of electric energy is made from medium (between 100°C and 200°C) and high- (>200°) temperature geothermal systems. Almost all the high-temperature fields exploited today are hydrothermal systems from which heat is extracted by means of wells producing fluids contained in a permeable reservoir. According to thermodynamic conditions, the reservoir can be either vapour- or liquid dominated depending on the fluid phase controlling the reservoir pressure distribution.

Geothermal power development requires a long project execution cycle, which the IGA (2014) guide divides into eight key phases: (1) preliminary survey; (2) exploration; (3) test drilling; (4) project review and planning; (5) field development; (6) power plant construction; (7) commissioning; and (8) operation. The three first phases (which could be broadly called the exploration stage) are seen as the highly risky part of the project development, because either confirm the existence of a geothermal reservoir suitable for power generation or not. According to Gehringer and Loksha (2012), it may take approximately seven years (usually between 5 and 10 years) to develop a typical full-size geothermal project with a 50 MW turbine as the first field development step. Therefore, it could not be regarded as a quick fix for any country's power supply problems, but rather should be part of a long-term electricity generation strategy.

CS9.2 METHOD

The technical report (Battistelli *et al.*, 2021) is the result of a desk-based work, consisted in a review of selected documents and news approximately from year 2005, searched on the web and dealing with geothermal resources development in East African countries. The review includes published papers, official reports, and documents that are available through the World Wide Web (WWW) mainly looking at institutional sites of involved stakeholders, both at the national and international levels.

The role and the activities performed by main international stakeholders have been summarized, highlighting both the results achieved in promoting the geothermal industry in East African countries and the needs for further efforts in specific fields.

As far as the status of geothermal energy development in each country is concerned, the analysis includes an overview of the energy sector in each country, with specific focus on the electric market and the present and planned resources used for electric power generation in order to highlight the possible role of geothermal resources utilization. In this context, the activities performed for geothermal resource exploration and development in each country are reviewed and summarized looking at the stated development plans of national stakeholders. Details on the activities performed and achieved results in studied geothermal prospects are given in order to highlight the actual status of each prospect and compare achieved results with stated development plans.

CS9.3 DISCUSSION

East Africa is characterized by the presence of the EARS with the Eastern branch, extending from Eritrea to Tanzania and crossing Djibouti, Ethiopia, and Kenya, and the Western branch extending from Uganda to Mozambique and crossing Burundi, Rwanda, Zambia, Tanzania, and Malawi (Figure CS9.1). While this geodynamic context creates high favourable conditions for the existence of geothermal resources at economically and technically drillable depths, at present only Kenya has developed its geothermal resources with an installed electric power of 865 MW, representing about 30% of total installed power, against estimated resources amounting to some 7000 MW.

Currently, it appears to be evident that the countries crossed by the Eastern Branch of EARS have a definitely higher geothermal potential, mainly concentrated in the Afar depression and the Ethiopian and Kenyan Rift Valleys. Even if not huge on an absolute scale, the resources inferred in Eritrea, Djibouti, and the Comoros (the latter not actually pertaining to the Eastern Branch of EARS), if developed, would almost satisfy their present and future electric network base loads. In the case of the countries crossed by the Western Branch, they have a lower geothermal potential, mostly related to medium, rarely high, temperature fault controlled geothermal systems, whose utilization for electric power generation would require ORC power plants. As a consequence, about 95% of EARS estimated potential amounting to some 22 400 MW belongs to geothermal areas located along the Eastern Branch.

The role of geothermal energy in the energy mix of the East African countries depends on the present status of the energy sector of each country, on the potential of indigenous energy sources, including geothermal energy, and strategic choices taken by each government. There are several reasons for the delay of geothermal resources development experienced so far by these countries, such as:

- Lack of clear and coherent legislative frameworks, regulating the activities of both public and private investors in several countries.
- Lack of local technical and managerial skills, able to conveniently support the exploration and exploitation of geothermal resources.
- The remoteness of many East Africa geothermal areas from developed O&G regions, where most of the drilling contractors and service providers are based, and then the absence of infrastructures and logistic facilities supporting the drilling activities characterizing well-developed O&G regions.
- Inadequate financing of the early stages of geothermal projects; commercial banks reluctance to participate in the exploration phase and the need for more risk reduction opportunities, which facilitate the investment by both public and private operators.
- Competition from other renewable energy sources, such as hydropower, solar, and wind, which creates a challenging environment for geothermal projects in the region.
- The issue of remunerative price for the generated electric power in still poorly developed in national electric markets.

To help East African countries to overcome the above issues, international organizations and financial institutions, such as WB, AU, EU, IRENA, NDF, AFD, AfDB, JICA, USAID, etc., are actively collaborating with national governments to create the necessary legislative framework in each country. They have facilitated the capacity building with the organization of dedicated courses and conferences

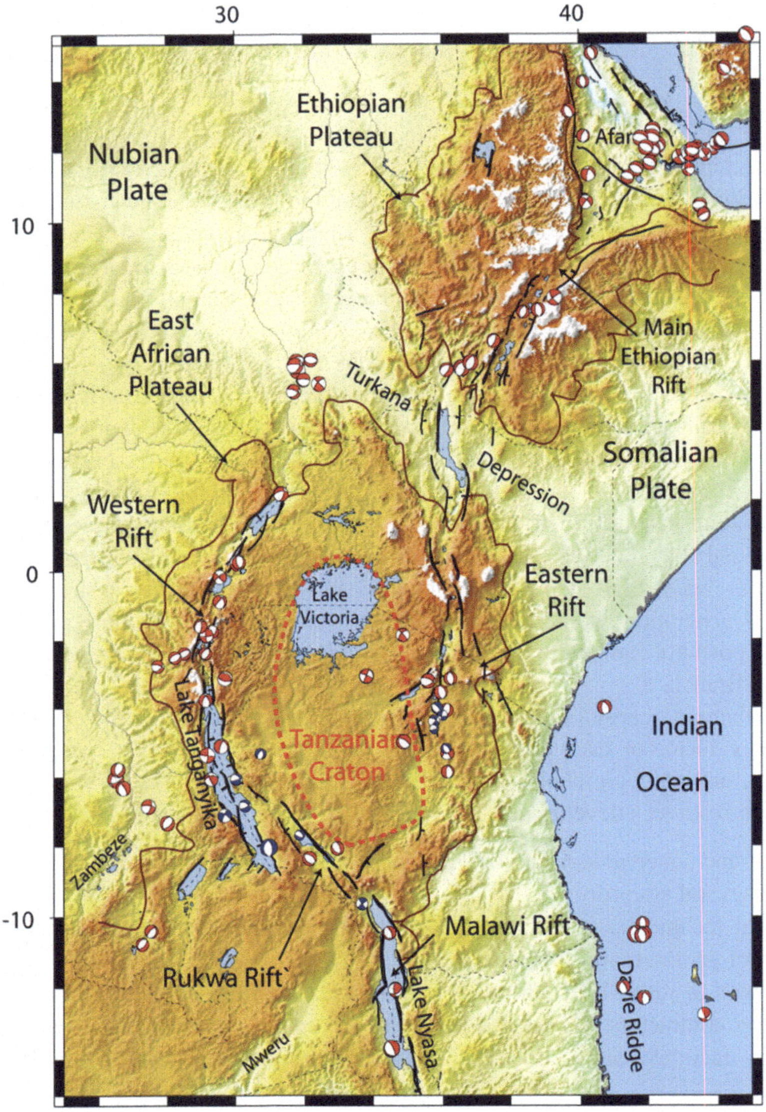

Figure CS9.1 The East African rift system (Omenda, 2018).

and the creation of the Africa Geothermal Centre of Excellence (AGCE) in Kenya, taking advantage of the existing training facilities of GDC and KenGen. On the other hand, financial and international institutions are providing both grants and low interest loans to help public and private operators in the various steps of geothermal resource development, from the exploration surveys to the construction of power plants.

In fact, geothermal power plant development involves substantial capital requirements due to exploration drilling costs, for which it can be difficult to obtain bank loans. Since geothermal exploration is considered high risk, developers generally need to obtain some type of public financing. This risk is derived from the fact that capital is required before confirmation of resource presence or exploitability, and therefore before project profitability can be determined.

Some of the Governments (Djibouti, Kenya, Uganda, Tanzania) have decided to reduce this risk and the cost of capital for private developers by creating public companies in charge of initial exploration activities and in some case also of exploitation of geothermal resources to provide private companies (that install power plants and supply electricity to their customers) with the required steam. An important risk mitigation opportunity is represented by the Geothermal Risk Mitigation Facility for Eastern Africa (GRMF), which is providing grants covering a variable costs fraction for infrastructure construction, surface exploration surveys and exploration drilling, the latter being the phase characterized by the higher mining risks. After five GRMF Application Rounds, grants have already been awarded to 30 projects located in Djibouti, Ethiopia, Kenia, Uganda, Tanzania, and Comoros. The opening of the expressions of interest for the 6th Application Round was done on 3 August 2020. In addition, several international stakeholders are actively supporting all the phases of geothermal field development, from exploration to power plant EPC, with grants and soft loans and providing technical assistance and consultant support to national institutions and geothermal operators.

Historically, reconnaissance and preliminary surface studies on geothermal prospects in East Africa were performed by public institutions or companies supported by international donors and consultants. Often, this approach has been characterized by a discontinuous performance of exploration phases separated by long periods of inactivity, sometime accompanied by the switch of operations from one institution to another one, with loss of skilled personnel and know-how. More recently, most of the countries have developed regulatory environments in which both public and private operators, as well as private–public initiatives, are allowed to develop the geothermal resources.

The example of Kenya, with an institutional setup of its energy sector similar to that of the most advanced geothermal countries in the world, testifies that the opening to private investors and operators, as well as to the collaboration between public companies in charge of the exploration and field management and independent power producers (IPP), allows an accelerated and more effective development. Other countries are following Kenya in establishing a clear regulatory environment and accelerating the initial prospects exploration by both dedicated public companies and private developers.

Regarding the forecasted role of geothermal energy in the generation of electricity, the scenarios developed by IEA (2019) (Figure CS9.2) show the electricity supply by type, source, and scenario in sub-Saharan Africa, excluding South Africa, in 2030 and 2040.

The situation in 2018 is compared to two different scenarios, namely the IEA's Stated Policies and the Africa Case, foreseen for year 2040. The IEA's Stated Policies scenario is based on current and announced policies, while the Africa Case scenario is a new scenario built by IEA around Africa's own vision for its future. It incorporates the policies needed to develop the continent's energy sector in a way that allows economies to grow strongly, sustainably, and inclusively.

In 2018 geothermal power accounted for 2% of electricity generation and is expected to represent in 2040 4% by both IEA's scenarios. The two scenarios foresee an increment of electric generation in 2040 of about 4 and 7.8 times, respectively, with respect to the present generation capacity, which implies an increment of geothermal generated electricity of about 8 and 15.6 times, respectively. Thus, geothermal energy is expected to double its contribution share in 2040, but still representing a small fraction of electricity generation, in particular if compared to the important increment of Solar PV, which will compensate for the reduction of hydropower contribution. On the other hand, both scenarios suggest that even if most of the investments on renewable energies will be drained by Solar PV, geothermal will anyway experience a large increment of generated energy and then of installed power.

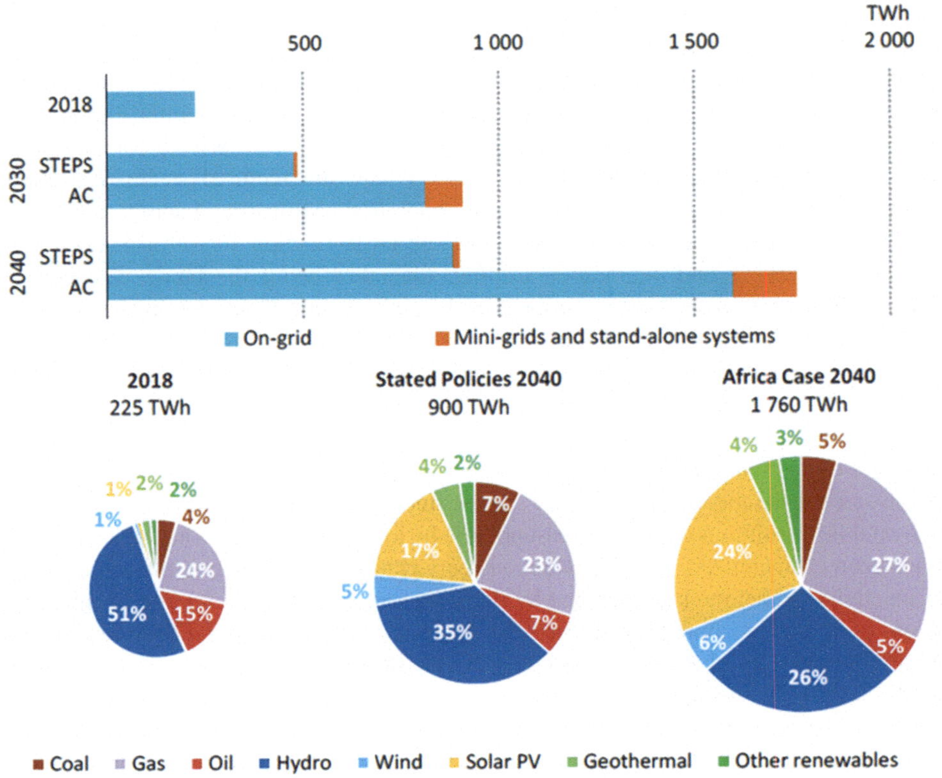

Figure CS9.2 Electricity supply by type, source and scenario in sub-Saharan Africa (excluding South Africa), in 2018, 2030 and 2040 (IEA, 2019).

CS9.4 CONCLUSIONS AND RECOMMENDATIONS

To help East African countries to overcome the identified barriers to the development of geothermal resources utilization, international organizations, and financial institutions are actively collaborating with national governments to create the necessary legislative framework in each country, to facilitate the capacity building with the creation of excellence centres and the organization of dedicated courses and conferences.

On the other hand, financial and international institutions, such as WB, AU, EU, IRENA, NDF, AFD, AfDB, JICA, USAID, etc., are providing both grants and low interest loans to help public and private operators in the various steps of geothermal resource development, from the exploration surveys to the construction of power plants.

In addition, the following technical approaches, derived from experiences and lessons learned, are believed to reduce risks and improve the bankability of geothermal projects (IRENA, 2018):

- Sound exploration for high-quality geological data.
- Linking technical and commercial analyses to the development of realistic prefeasibility studies prior to making major investments.
- Generating early revenues through wellhead generators. Actually, installing wellhead power plants is advantageous when an early electric generation can be obtained during a long-term field development

in quite large fields, and when the wellhead power plants can be relocated on another field or field sector when the final power plant starts its operations.

- Supplement project revenues through direct use applications and sale of other by-products such as lithium, CO_2, silica, etc.

In any case, geothermal energy shall be competitive in relation to other energy sources, either other renewables or fossil fuels. According to IRENA (2019), most of the plants allows a levelized cost of electricity (LCOE) lower than about 0.08 USD/kWh, which is competitive with electricity generated with fossil fuels (Figure CS9.3). It is reasonable, that cheaper renewable energy sources like solar and wind not affected by the mining risks of geothermal energy may likely be preferred by many international and national investors. The important foreseen development of electric energy market in East African countries allows anyway to expect a corresponding development of geothermal resources to diversify the use of indigenous renewable resources and take advantage of the peculiar advantages of geothermal energy generation (stable production, low GHG emissions, low operating and maintenance costs).

The countries that at present show the best geothermal perspectives, mainly located along the Eastern Branch of the EARS, are Djibouti, Ethiopia, Kenya, and Tanzania, while the limited resources inferred in Eritrea and the Comoros would anyway be able to cover the present electric base load in both countries.

In conclusion, thanks to the efforts of both national governments and international stakeholders, the geothermal energy in Eastern Branch countries of EARS seems to be at a turning point in particular in Ethiopia and Djibouti, with Kenya going on in an accelerated way along an already established

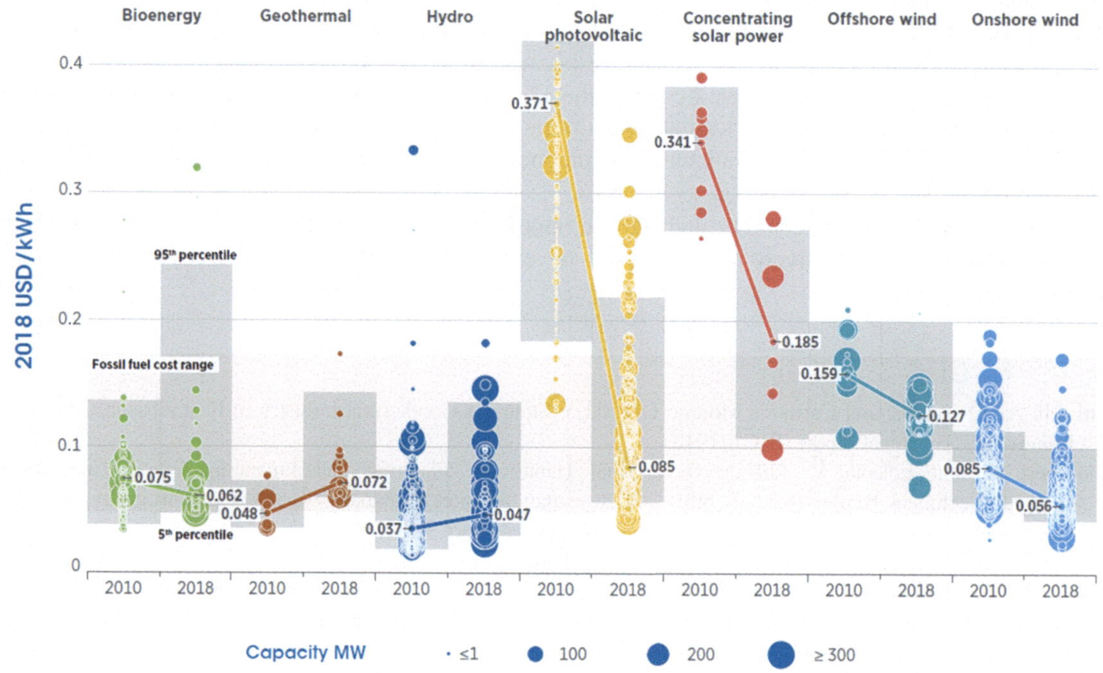

Figure CS9.3 Global LCOE evolution of utility-scale renewable power generation technologies in the 2010–2018 period. Real weighted average cost of capital is 7.5% for OECD countries and China and 10% for the rest of the world (IRENA, 2019).

successful path. On-going exploration in the Comoros has also good perspectives with geothermal potential to be confirmed, but largely exceeding the present base load of the country.

The geological settings and the exploration activities performed so far suggest that the countries crossed by the Western Branch of EARS have a lower geothermal potential, mostly related to medium, rarely high, temperature fault controlled geothermal systems whose utilization for electric power generation would require ORC power plants. Experiences recently gained with the exploration of fault-controlled systems in the Western Branch and related new achieved understanding, have implications for both tailored geological exploration approaches and the identification and prioritization of prospects in the Western Branch countries, which will likely allow to identify new promising prospects.

CS9.5 STUDY LIMITATIONS AND THE WAY FORWARD

The focus of the study was on geothermal activities aimed at generating electric power by using either flashing or ORC plants with geothermal fluids extracted from medium-to-high temperature hydrothermal systems. Thus, direct uses of geothermal energy such spas, cooking, space heating and cooling, greenhouse heating, crop drying, aquaculture and heat for industrial processes, as well as the possible selling of by-products (lithium, silica, CO_2) were not addressed in the related report. Direct uses represent the natural utilization of low-temperature resources, but can also complement the exploitation of medium and high-temperature resources for electric power generation to improve the heat recovery and the project rentability.

The final report is the result of a desk-based work, consisted in a literature review of selected papers and news approximately from year 2005, searched on web resources. The review includes published papers, reports, and documents that are available through the World Wide Web (WWW). While any reasonable effort has been assured to collect the relevant information within the time constraints of the study, of course the literature review cannot be exhaustive because of several projects underway and also many international and national stakeholders acting in the 11 African countries considered. In addition, while most of the general information is available to the public through the WWW, the details on specific initiatives are often not readily available and, on the other hand, the published information is not always updated.

The outcomes of the study could be updated and improved by contacting the various stakeholders, both national and international, actively involved in the development of geothermal resources in the countries crossed by the EARS.

REFERENCES

Battistelli A., Crestaz E. and Carmona-Moreno C. (2021). Status of Geothermal Industry in East African Countries. Report JRC Technical report (JRC121913).

Gehringer M. and Loksha V. (2012). Geothermal Handbook: Planning and Financing Power Generation. ESMAP Technical Report 002/12. https://www.esmap.org/sites/esmap.org/files/DocumentLibrary/FINAL_Geothermal%20Handbook_TR002-12_Reduced.pdf (last accessed July 2021).

IEA (2019). Africa Energy Outlook 2019. World Energy Outlook Special report.

IGA (2014). Best practices guide for geothermal exploration. IGA Service GmbH in partnership with IFC.

IRENA (2018). Geothermal finance and risk mitigation in east Africa. Key findings from a regional workshop, Nairobi, Kenya, 31 January–2 February, 2018.

IRENA (2019). Renewable Power Generation Costs in 2018, International Renewable Energy Agency, Abu Dhabi. ISBN 978-92-9260-126-3.

Omenda P. (2018). Update on the status of geothermal development in Africa. In: Proceedings of GRC Annual Meeting and Expo. Reno, Nevada, USA: GRC.

Water–Energy–Food (WEF) governance and institutions in the Southern African Development Community (SADC)

A. Takawira[1], J. K. Tiruneh[1], P. K. Kabeya[2], M. Ntlamelle[2] and D. Samikwa[2]

[1]GWP Southern Africa, Pretoria, South Africa
[2]Southern African Development Community (SADC) Secretariat

ABSTRACT

The Southern African Development Community (SADC) region faces water, energy, and food (WEF) insecurity, even more in view of expected increased stress due to climate change. The opportunities of the WEFE Nexus approach to effectively achieve sustainable development and moving forward towards attaining most of the SDGs targets are discussed. The WEFE Nexus addresses the challenge of sectorial management of resources through the adoption of harmonized institutions and policies, as well as setting targets and indicators to implement and assess resource management for sustainability. SADC is embracing the WEF Nexus approach providing good political and policy support, opening up dialogues in establishing clear directions and establishing regional framework that will create an enabling environment for Nexus approaches and facilitate nexus investment in the region. The SADC WEF Nexus Regional Framework is expected to bring coherence between the WEF policies; facilitate institutional coordination; align development strategies, targets, and programmes of the three sectors; manage trade-offs; and promoting Nexus investments in the region.

Keywords: WEF Nexus, governance, SADC region

CS10.1 BACKGROUND

Water, energy, and food (WEF) resources are vital for human wellbeing, poverty reduction, and sustainable development. The Southern African region represents a wide range of resource and climate contexts with varied supplies of water, food, and energy. About 60% of the population of the Southern African Development Community (SADC) live in rural areas relying on rain-fed agriculture, lacking basic services of clean energy, water, and sanitation, yet the region is endowed with vast natural resources. Ensuring WEF security has dominated the development agenda of southern African countries, centred on improving livelihoods, building resilience, and regional integration. Increasing demands for water, land, and energy resources due to population growth, increasing urbanization, and increasing economic growth are the major challenge in the region. This challenge is exacerbated by climate change. This is particularly concerning for the SADC region due to dependence on climate-sensitive sectors of agriculture and energy, which heavily depend on water resources.

The water–energy–food (WEF) Nexus approach has potential application in the region for ensuring security of WEF; and for bringing resource use efficiency. It provides opportunity to stabilize competing demands and promote regional integration, particularly in the SADC where resources are mostly transboundary. The WEF Nexus approach can help to ensure that development of one of the sectors has minimum impacts on the other. Sectorial collaboration is particularly relevant in SADC as watercourses and electricity grids are shared among countries.

This article attempts to review the existing governance and institutions for managing water, energy, and food sectors in the SADC region, and identifies potential areas for change in strengthening the WEF Nexus governance system of the region.

CS10.2 WEF RESOURCE ENDOWMENTS IN THE SADC REGION

Water resources significantly vary in distribution, availability, and usage across the SADC region. Approximately 75% of the SADC region, most of which in the southern part, is semi-arid to arid receiving less than 650 mm of rainfall per annum. The rest, 25%, which is mostly occupied by northern countries that are closer to the Equator, is classified as sub-humid receiving between 651 and 2500 mm of rainfall per annum.

According to the SADC (2012), there is an estimated total of 2300 km^3/year of renewable water resources available to the SADC region's population of 260 million people. The level of abstraction is only 44 km^3/year or 170 m^3/capita/year. Of the 44 km^3/year abstracted, 77% is used for irrigation, 18% for domestic purposes while 5% is used by industry.

Seventy per cent of surface water resources are in 15 transboundary river basins (Figures CS10.1, CS10.2 and Table CS10.1).

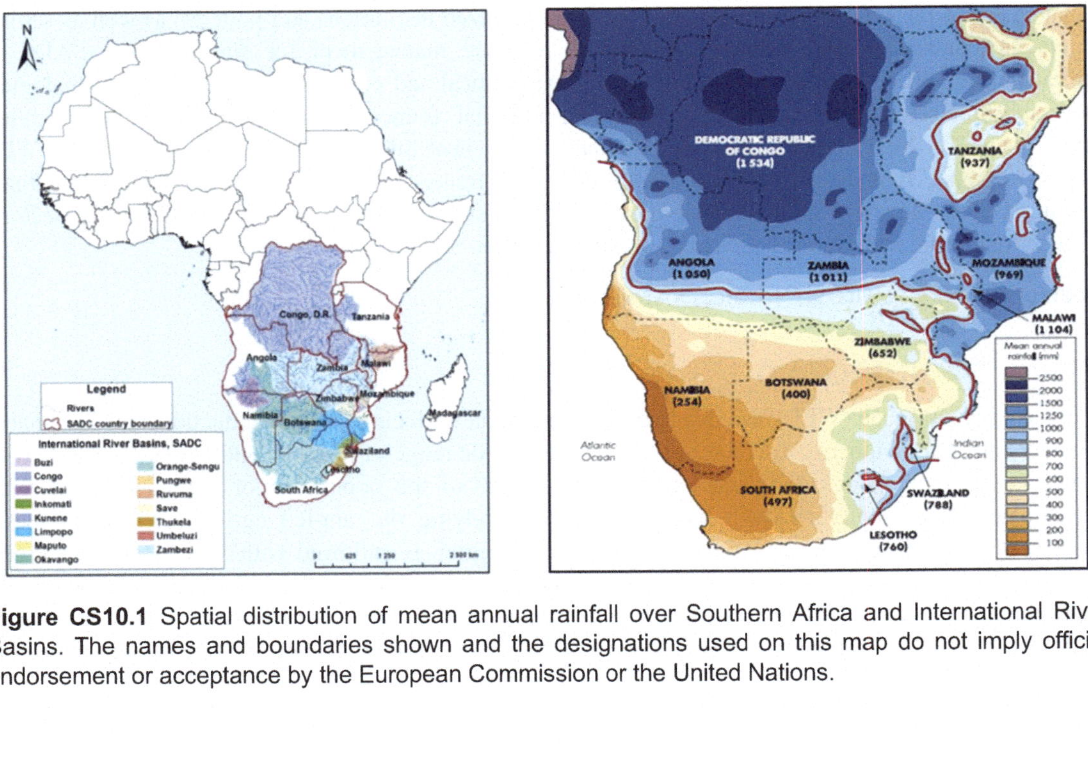

Figure CS10.1 Spatial distribution of mean annual rainfall over Southern Africa and International River Basins. The names and boundaries shown and the designations used on this map do not imply official endorsement or acceptance by the European Commission or the United Nations.

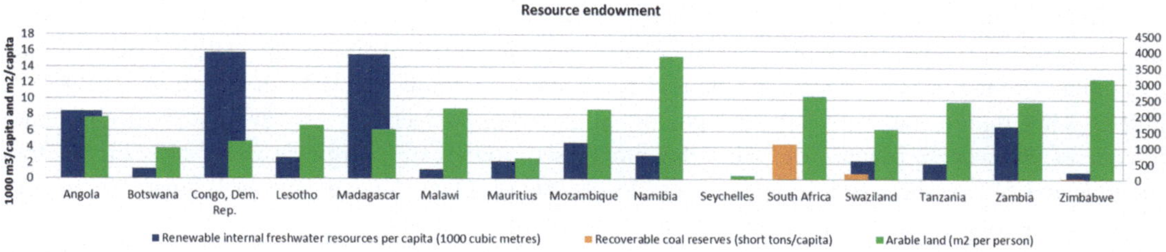

Figure CS10.2 SADC Resource Endowment (*Source*: SADC 2012 (RIDMP)).

The transboundary nature of the river basins signifies the importance of watercourses in promoting regional integration and development. For example, five SADC countries have water resources dependency ratios of over 50%, that is, they rely on water generated outside their borders to supply more than half of their total water requirements (Malzbender & Earle, 2009).

Although the southern parts are generally drier, the Congo, Zambezi, and Orange–Senqu basins have the potential to generate significant regional benefits through water transfer and hydropower generation.

The uneven distribution of resources in the region creates uneven demand pressures on raw materials and natural resources. As a result, demand for water and energy resources is evidently concentrated in the southern parts of the region. In contrast, the northern parts of the region (e.g., the Congo and Zambezi river basins) are endowed with the abundant water resources that could sustainably deliver these inputs.

Table CS10.1 SADC transboundary river basins (IRBs) and the riparian states.

River Basin	Riparian States	Area (km^2)
Buzi	Mozambique, Zimbabwe	27,000
Congo	Angola, Democratic Republic of Congo, Tanzania, Zambia	3800,100
Cuvelai	Angola, Namibia	167,000
Incomati	Mozambique, South Africa, Eswatini	46,740
Kunene	Angola, Namibia	106,560
Limpopo	Botswana, Mozambique, South Africa, Zimbabwe	408,000
Maputo	Mozambique, South Africa, Eswatini	29,970
Nile	Democratic Republic of Congo, Tanzania	3,200,000
Okavango	Angola, Botswana, Namibia	323,192
Orange	Botswana, Lesotho, Namibia, South Africa	1,000,000
Pungwe	Mozambique, Zimbabwe	31,000
Ruvuma	Malawi, Mozambique, Tanzania	152,000
Save	Mozambique, Zimbabwe	115,700
Umbeluzi	Mozambique, Eswatini	10,900
Zambezi	Angola, Botswana, Malawi, Mozambique, Namibia, Tanzania, Zambia, Zimbabwe	1,570,000

The SADC region is endowed with vast energy resources, although availability varies from country to country. There is huge potential of hydropower generation in Angola, the DRC, Mozambique, and Zambia. The region's hydropower potential is estimated at about 150 GW, of which only 12 GW is installed. The region currently shares power grids whose electricity is generated from shared watercourses. However, biomass remains the most used source of energy as only 24% of the total population and 5% of rural people have access to electricity. Over dependency on biomass energy has contributed to massive deforestation and land degradation in the region.

Demand for energy in the region continues to increase due to population and industrial growth and urbanization. According to the SADC Regional Infrastructure Development Master Plan (RIDMP) of 2012, assuming an average economic growth rate of 8% per annum, energy demand is expected to increase to more than 77 GW by 2020 and to over 115 GW by 2030, exerting more pressure on water resources (SADC, 2012).

Agriculture is the main catalyst for economic development in the SADC as more than 60% of inhabitants depend on it for their livelihoods, providing their subsistence, employment and income. The performance of the agriculture sector, therefore significantly impacts on economic growth, poverty reduction, and food security. Despite its importance, performance of the sector has been insufficient to significantly contribute to regional economic growth and address food and nutrition security issues in the region.

Agriculture is mainly rain-fed. There is about 50 million hectares of irrigable land available within the SADC region, of which only 3.4 million hectares (7%) is irrigated (SADC, 2011).

The region's plans for increasing agricultural productivity are underpinned on increasing land area under cultivation, and under irrigation. This alludes to land-use changes in some cases, and increasing the amount of water withdrawals and associated energy outlays needed for irrigation. Proponents argue that this is feasible given the region's large tracts of underutilized arable land and water resources (dams). On the other hand, there is an argument that much of the land is degraded and may not be suitable for agriculture, and that in some countries such as South Africa, most of the available water is already allocated and with little scope for building new dams.

In summary, considering the uneven distribution of water, energy, and land resources in the SADC region and the overall increasing demand for such resources, it becomes evident that a regional WEF Nexus approach has potential to increase benefits and reduce risks. There is considerable potential for coordinated infrastructure investment to improve overall use of WEF Nexus resources in the SADC.

CS10.3 WEF NEXUS-RELATED INSTITUTIONS AND POLICIES OF THE SADC

Due to the shared resources and common climatic, cultural, and political history, the region has put in place institutions, policies, and other frameworks to oversee and direct water, energy, and agriculture resources at regional level. However, coordination among the policies and institutions is inadequate.

The SADC Industrialisation Strategy and Roadmap (2015–2063) was approved in April 2015. This framework is aimed at achieving industrial development and has been placed at the core of the developmental integration agenda of SADC. Inherent in this policy is recognition from SADC member states that industrial development is central to diversification of their economies; development of productive capacity; and the creation of employment in order to reduce poverty and set their economies on a more sustainable growth path. WEF are key priority areas for SADC. Integrated water, food, and energy planning and promoting regional cooperation are considered a strategy to meet socio-economic security targets, and improve natural resource use efficiencies in the region.

CS10.3.1 WEF-related policies

The SADC region has a number of policies that are related to developing and managing water, energy, and land resources.

- The SADC Regional Water Policy was adopted in 2005. The Policy is implemented through a Regional Strategic Action Plan (RSAP), a 5-year Regional Water Programme that aims to achieve an equitable and sustainable utilization of water for social and environmental justice, regional integration and economic benefit for present and future generations. The current RSAP IV (2016–2020) emphasizes the importance of infrastructure development and water resource management for food security in the water–food nexus, and the stronger urgency to take action in the view of climate variability and change. The action plan recognizes the role of the nexus in adapting to the challenges posed by population growth and climate change and variability, as well as in optimizing resource use in order to achieve regional goals and targets.
- The SADC Protocol on Shared Watercourses (SADC, 2000 (revised)), fosters closer cooperation for judicious, sustainable, and coordinated management, protection and utilization of shared watercourses, and advance SADC's agenda of regional integration and poverty alleviation. As a result, most shared river basins have basin level agreements in place which oversee the day-to-day management of the basins with assistance from the SADC Water Division.
- The SADC Protocol on Energy (SADC, 1996), highlights the development and updating of a regional electricity master plan, the development, and utilization of electricity in an environmentally sound manner, and emphasizing the need for universal access to affordable and quality services.
- A Regional Energy Access Strategic Action Plan (REASAP) was approved in 2011, setting broad goals for improving access to modern forms of energy as well as specific policy mechanisms to achieve increased access. However, there is no Regional Strategic Action Plan for Energy that has been officially adopted to-date. A Renewable Energy Strategy and Action Plan (RESAP) was approved in 2016, and a SADC Centre for Renewable Energy and Energy Efficiency (SACREEE) has been established. SADC also has energy cooperation policy and strategy (1996).
- The SADC Regional Agricultural Policy (RAP) (SADC, 2014) envisages integrated approaches on water resources management and emphasizes the importance of improving agriculture performance to meet the food and water security as well as attain sustainable economic development objectives at the regional level. The SADC's Regional Agricultural Investment Plan is derived from the Africa-wide Comprehensive Africa Agricultural Development Programme (CAADP) that promotes the doubling of irrigated area from 3.5% to 7% by 2025. The CAADP (CAADP, 2009) provides a common framework for stimulating and guiding national, regional, and continental initiatives on enhanced agricultural productivity and food security. SADC has a 10 years food and nutrition strategy (2015–2025).

CS10.3.2 WEF Nexus-related institutions

The regional level institutional structures for managing the water, energy, and agriculture sectors are summarized in Table CS10.2.

CS10.4 SADC WEF NEXUS GOVERNANCE AND INSTITUTIONAL ISSUES

CS10.4.1 Uncoordinated sectorial plans and targets

The Regional Infrastructure Development Master Plan (RIDMP) (SADC, 2012) sets out the region's infrastructure development targets as shown in Table CS10.3 (Figure CS10.3).

Table CS10.2 Regional level institutional structures for managing the water, energy, and agriculture sectors.

Sector	Regional Structures	Objectives/mandates
Food, Agriculture, Natural Resources	Food, Agriculture and Natural Resources (FANR)	• Development, coordination, and harmonization of agricultural policies and programmes
Water and Energy	Infrastructure and Services Directorate	• Development, coordination, and harmonization of energy, transport and communications, tourism and water policies, strategies, programmes and projects; • Coordination and promotion of integrated management of trans-boundary water, tourism, transport and communication and energy resources for regional integration and development;
	Water Division SADC-RBOs	• To oversee harmonization of national water use policies, and moderate transboundary issues.
	Southern African Power Pool (SAPP)	• Enhance regional cooperation in power development and trade, and to provide non-binding regional master plans to guide electricity generation and transmission infrastructure delivery.
	SADC Centre for Renewable Energy and Energy Efficiency (SACREEE)	• Promote renewable energy and energy efficiency technologies • Develop sound policy, regulatory, and legal frameworks, and build capacities

The RIDMP ambitious sectorial targets do not seem to consider the available water, land, and energy resources of the region.

CS10.4.2 Inadequate inter-sectorial dialogues

SADC provided WEF Nexus regional platforms, mainly based on the existing SADC multi-stakeholders water dialogues. The 2013 SADC Water Dialogue focused in exploring the WEF Nexus. The discussion was around infrastructure development and strengthening institutions for economic growth. The dialogue

Table CS10.3 SADC WEF-related targets.

Project	Potential	Baseline (2012)	Targeted Plan (2027)
Hydropower	150 GW	12 GW	Increase to 75 GW (50% of potential)
Irrigation	50 M has	3.4 M has	Increase to 10 M has (13% of potential)
Annual Renewable WR	2300 km^3	14% retained	Increase to 25%
Access to clean water		61%	Increase to 75%
Access to sanitation		39%	Increase to 75%

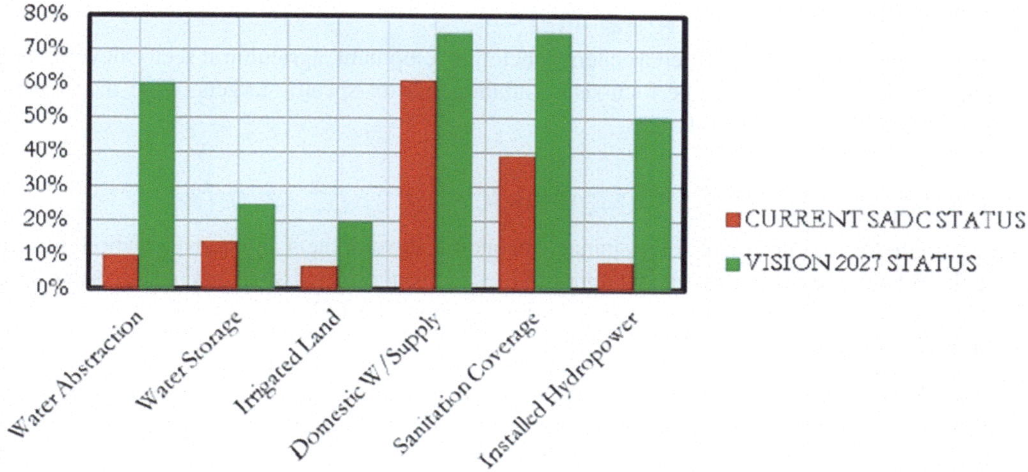

Figure CS10.3 SADC regional infrastructure development targets (2013–2027) (*Source*: SADC 2012).

recognized the need for integrated planning and implementation of development programmes; and recognized that the Nexus approach can provide opportunity for coherent and well-planned development and use of water, energy and food resources. Other dialogues include the 2016 SADC Ministerial workshop on energy and water crisis, and the 2017 SADC Water and Energy Ministers Conference. The 2017 SADC multi-stakeholders water dialogue (November 2017) will focus in fostering regional value chains and job creation through the WEF Nexus approaches.

The key issues that emerged from the SADC Nexus Dialogues include:

- Enhance sectorial collaboration and joint management in the region
- Promote a nexus thinking at all levels (regional, transboundary, and national)
- Anchor Nexus in the sustainable infrastructure development and industrialization agenda
- Develop an operational framework to drive and guide sectorial collaboration in implementing priority investment projects

CS10.4.3 Sector-focused policies and institutions with inadequate coordination mechanism

The WEF Nexus could lead to more optimal allocation of resources, promotion of inclusive and sustainable regional economic growth. However, this will depend on the availability of support from harmonized WEF institutions and policies.

The water, energy, and food policies in the SADC region are sector-focused with limited recognition of the interlinkages between the water–land–energy resources. Similarly, the institutions and governance arrangements are also structured around the sectors without looking at the interlinkages. There is a challenge in shifting from nationally driven agendas to regionally driven agendas due to the transboundary nature of resources that could ensure regional WEF.

CS10.4.4 Non-coordinated programmes

Programmes are more sector-based such as energy sector development, agricultural sector development or water supply service programmes. The focus is on attaining sector-specific targets rather than meeting a comprehensive and integrated WEF targets.

CS10.5 THE WEF NEXUS FRAMEWORK FOR SADC REGION

WEF are central to the region's plans for sustainable economic development and transformation. It is in this regard that the WEF Nexus offers significant opportunities for a coordinated approach to addressing some of the region's pressing challenges and achieving regional goals. The following specific benefits could be realized through adoption of a WEF Nexus approach:

- *Regional integration.* The WEF Nexus provides a meaningful platform for coordinated access, utilization and beneficiation of shared resources and potential for effective synergies and trade-offs between the WEF Nexus components. The WEF Nexus also provides an opportunity to harmonize existing institutions and policies and translate them into coordinated balanced strategies that can contribute towards inclusive development, socio-economic security, and regional integration.
- *Sustainable economic development.* The WEF Nexus promotes the inseparable link between the use of resources to provide basic and universal rights to food, water, and energy security that will promote sustainable and inclusive economic development.
- *Harmonization of institutions and policies.* Harmonizing institutions and policies among the three sectors minimizes cross-sectorial conflicts, maximizes synergies, mitigates trade-offs, reduces implementation costs, and achieves policy objectives through a systems approach. Harmonized policies ensure systematic promotion of mutually reinforcing strategies and instruments and resolve policy conflicts in order to meet the competing demands for resources.
- *Build resilience.* The WEF Nexus approach provides opportunity for increasing regional resilience against climate change impacts and mitigating vulnerabilities through coordinated WEF infrastructure development, improved management of transboundary natural resources, maximizing on regional comparative advantages for agricultural production and unlocking more resources for climate proofing through increased efficiencies.
- Promote investment in infrastructure development. The WEF Nexus promotes investment.

For the SADC region to reap the benefits of the nexus approach, it needs to establish a regional WEF Nexus Framework. Preliminary level assessment and consultations with stakeholders indicated that the regional framework are represented in Figure CS10.4.

CS10.6 CONCLUSIONS AND RECOMMENDATIONS

The SADC region faces WEF insecurity. Climate change projections suggest increased stresses on the WEF sectors, thus challenging future development plans. The WEF Nexus approach offers opportunities to effectively achieve sustainable development through cross-sectorial collaboration and harmonization of sectorial policies. Adoption of the WEF Nexus approach would be a step forward towards attaining most of the SDGs targets. As the vast and unexploited resources within the region are shared, the WEF Nexus presents opportunities for regional integration, coordinated resources development, resilience building, and reduction of vulnerabilities and attainment of regional development targets.

The WEF Nexus addresses the challenge of sectorial management of resources through the adoption of harmonized institutions and policies, as well as setting targets and indicators to implement and assess

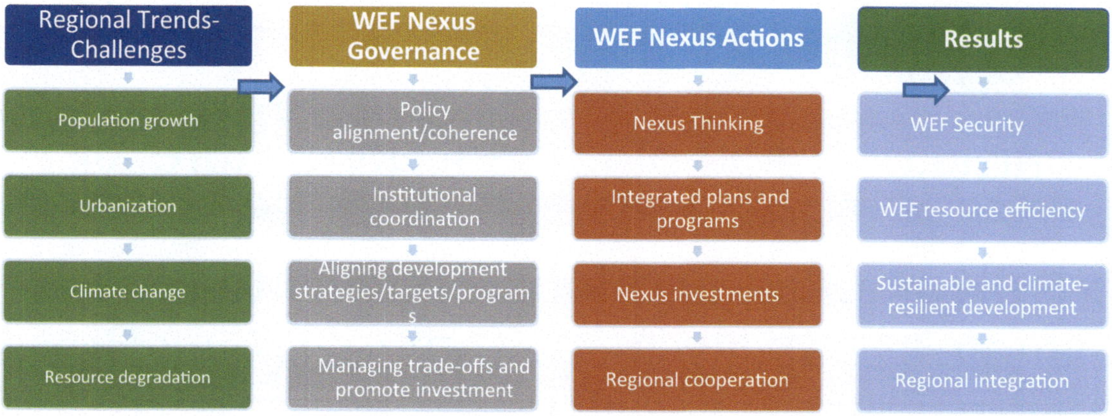

Figure CS10.4 Preliminary level assessment and consultations with stakeholders.

resource management for sustainability. In the SADC, the WEF Nexus could prove to be valuable by promoting inclusive development and transforming vulnerable communities into resilient communities.

SADC is embracing the WEF Nexus approach providing good political/policy support, opening up dialogues in establishing clear directions and establishing regional framework that will create an enabling environment for Nexus approaches and facilitate nexus investment in the region.

The SADC WEF Nexus Regional Framework is expected to bring about alignment/coherence between the WEF policies; facilitate institutional coordination; align development strategies/targets/programmes of the three sectors; and manage trade-offs and promote Nexus investments in the region.

REFERENCES

Malzbender D. and Earle A. (2009). Water Resources of SADC: Demands, Dependencies, and Governance Responses. In: Rethinking Natural Resources in Southern Africa, M. Pressend and T. Othieno (eds.), Institute for Global Dialogue, Midrand, pp. 85–106.

Regional Infrastructure Development Master Plan (RIDMP) (August 2012). In https://www.sadc.int/files/7513/5293/3530/Regional_Infrastructure_Development_Master_Plan_Executive_Summary.pdf (accessed 3 January 2019).

SADC Industrialisation Strategy and Roadmap (2015–2063) (29 April 2015). https://www.ilo.org/wcmsp5/groups/public/—africa/—ro-addis_ababa/—ilo-pretoria/documents/meetingdocument/wcms_391013.pdf (accessed 3 January 2019).

SADC Protocol on Energy (1996). In https://www.sadc.int/documents-publications/show/Protocol_on_Energy1996.pdf (accessed 3 January 2019).

SADC Protocol on Shared Water Courses (2000). In https://www.internationalwaterlaw.org/documents/regionaldocs/Revised-SADC-SharedWatercourse-Protocol-2000.pdf (accessed 3 January 2019).

SADC Regional Agricultural Policy (RAP) (2014). In http://www.inter-reseaux.org/IMG/pdf/Regional_Agricultural_Policy_SADC.pdf (accessed 3 January 2019).

SADC Regional Water Policy (August 2005). In http://nepadwatercoe.org/wp-content/uploads/Regional_Water_Policy.pdf (accessed 3 January 2019).

SADC STATISTICS YEARBOOK (2012). In https://www.sadc.int/information-services/sadc-statistics/sadc-statist/ (accessed 3 January 2019).

The Comprehensive Africa Agriculture Development Programme (CAADP) (January 2009). In http://www.resakss.org/sites/default/files/pdfs/comprehensive-africa-agriculture-development-progr-39420.pdf (accessed 3 January 2019).

Characterization of current agriculture activities, future potential irrigation developments, and food security to face climate variability in the Zambezi River Basin

A. Senzanje[1] *and T. L. Dirwai*[1,2]

[1]*Bioresources Engineering Programme and Centre for Water Resources Research, School of Engineering, University of KwaZulu-Natal, Pietermaritzburg, South Africa*
[2]*Soil-, Crop-, and Climate Sciences, University of the Free State, P.O Box 339, Bloemfontein, South Africa*

ABSTRACT

Water and agriculture aspects focusing on irrigated and rainfed agriculture through appropriate agricultural water management practices were analysed in the transboundary Zambezi River Basin (ZRB). This basin, the fourth largest in Africa, is facing many challenges from the WEFE Nexus perspective. Agriculture is the largest water consumer in the basin with more than 90% of the agricultural activity being based on flood plain cultivation and rainfed agriculture, and sustaining the bulk of the rural population. Irrigation is important in the basin, but on a comparative basis estimates range from 147,000 ha to 259,000 ha only, but because of its water-use intensity, it factors significantly in the water demand in the basin. Irrigation is estimated to consume 3235 million cubic meters of water currently amounting to 1.4% of the basin's renewable water resources. There is huge irrigation development potential in the basin, ambitious plans are afoot to triple the area under irrigation by 2025, which will increase the water for irrigation to 4.1% of the renewable water resources. Smallholder irrigation practices are dominant in the ZRB, consequently basic agricultural water management coupled with sustainable agricultural intensification is a key aspect of agricultural production. Typical practices in the basin include; gravity fed off-river and reservoir irrigation, dambo irrigation, motorized pumping irrigation, drip irrigation, including drip kits, sprinkler irrigation, and centre pivot irrigation. Since the ZRB is transboundary and there is competition for natural resources by sector (water, energy, agriculture) and by country (ZRB riparian countries), the WEFE Nexus represents a viable tool for resources management. An exploratory WEFE Nexus analysis of the ZRB was conducted based on selected water, energy, food, and ecosystem goods and services indicators. Most of the WEFE indicators showed marginal sustainability.

Keywords: agricultural water management, irrigation development, renewable water resources, smallholder irrigation, WEFE Nexus

CS11.1 INTRODUCTION

The Zambezi River Basin (ZRB) is the fourth largest river in Africa after the Congo, Nile, and Niger River basins. It is located in Southern Africa and is coordinated between 9°00′S to 20°30′S latitude and 18°20′E to 36°25′E longitude (Figure CS11.1) (Schleiss & Matos, 2010). The ZRB covers an area of 1.4 million km^2 and stretches approximately 2600 km. The basin is shared by eight riparian countries (Angola, Botswana, Malawi, Mozambique, Namibia, United Republic of Tanzania, Zambia, and Zimbabwe), with Zambia having the largest share (41%), followed by Angola and Zimbabwe which have slightly less than half the Zambian portion (18.5 and 15.6%, respectively) (Schleiss & Matos, 2010). Three of the countries (Republic of Tanzania, Namibia, and Botswana) have less than 2% of the river basin each. The ZRB is commonly split into three main regions: Upper Zambezi, Middle Zambezi, and Lower Zambezi.

The average annual rainfall over the whole river basin is estimated to be 990 mm while the average annual evaporation is about 870 mm. The part of the ZRB which receives the highest rainfall (over 2000 mm per annum), lies in Tanzania (Cai *et al.*, 2017). Land cover in the ZRB consists of rainfed farming, forest, bushland, grassland, open water, and irrigated land. Forest and bushland take up 75% of land cover. Within the remainder 25%, rainfed agriculture occupies an estimated 13.2% of the land holding. Grassland and irrigated agriculture occupy 7.7 and 1.3%, respectively (Euroconsult Mott McDonald, 2007). Agricultural activities have been the main driver of land cover changes. According to Gomo *et al.* (2018), approximately 16% of natural forests have been converted to crop area over the past decade.

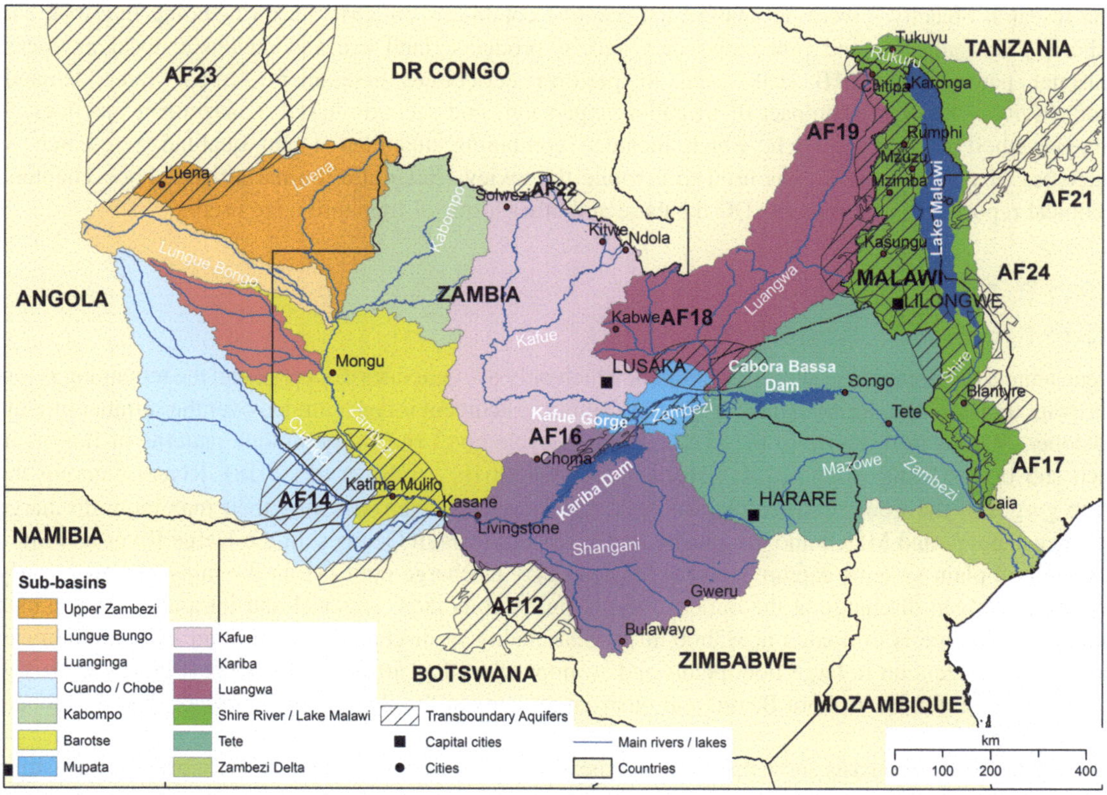

Figure CS11.1 The ZRB subbasins and the shared aquifers (*Source*: Cai *et al.*, 2017).

Agriculture is the largest man-made land use around the ZRB. It is a large contributor to GDP of riparian countries and livelihoods of its inhabitants. Malawi has the largest cultivated area within the ZRB (≈2 million hectares), followed by Zimbabwe and Zambia Angola, Botswana, and Mozambique have the least area under cultivation within the basin (<0.1 million hectares). With respect to renewable water resources, agriculture is the largest user in all riparian countries with countries such as Zimbabwe, Zambia, Tanzania, Mozambique, and Malawi having more than 75% of renewable water resources being consumed by agriculture (World Bank, 2008). Despite this, there is not much information on agricultural water management (AWM) in the basin at different scales (Manzungu *et al.*, 2017).

With respect to water and agriculture in the ZRB, questions arise include; how much land is available in the basin and of that how much is suited to agriculture and is being used thus, what typologies of agricultural practices exist in the basin and what are the key factors driving these, how much water is available in the basin and of that how much is allocated or available for agriculture, how much irrigation is taking place in the basin and what is the potential for further expansion with what water resources, what are the levels of agricultural water productivity in the basin, what options exist to improve agricultural water management, can there be trade-offs between rain fed and irrigated agriculture, and can the basin be eventually food secure? Within the context of the WEFE Nexus, the questions are, can this be used as an approach or tool to better manage resources in the basin for sustainable energy and food production?

CS11.2 METHODS

The research objectives were; to analyse the baseline conditions on agriculture (including livestock and fisheries) by gathering and processing data and by-products (land use and coverage, local practices, seasonal patterns) at ZRB scale; and to perform agriculture assessment (crops water demand, productivity and potential impact of irrigation expansion) and scenario-based management practices. A mixed-method review approach, which included combining quantitative and qualitative research or outcomes of process studies was used to compile the review. Scientific journal articles, book chapters, technical reports, dissertations, SADC database and other forms of literature were used.

CS11.3 DISCUSSION

CS11.3.1 Surface water

Mean annual precipitation is about 1000 mm of which only 8% generates discharge and the remaining is lost via evapotranspiration. Rainfall throughout the Zambezi catchment is concentrated over the summer months (October–March) in response to the ITCZ. The rain cycle gives rise to the unique patterns of run-off in each sub basin (Kling *et al.*, 2014; Schleiss & Matos, 2010; Zimba *et al.*, 2018). Rivers draining the steep gorges of the Central Africa plateau peak rapidly with the rain, reaching their maximum discharge between January and March and decreasing to dry season flows in October. In the Kafue River and Shire basin, flood plain systems capture flood water and delay discharge until late in the rainy season or early dry season. Mean discharge at the outlet of the basin exhibits large seasonal and intra-annual variations though its average is estimated at ≈3600 m^3/s. Seasonality in discharge is controlled by seasonality in precipitation, retention in large floodplains and swamps as well as artificial reservoirs (Pinay, 1988). The construction of Kariba, Cahora Bassa, and other large dams in the Zambezi system has altered Zambezi runoff pattern.

Climate change forecasts show that ZRB will be affected by climate change, with runoff being sensitive to variations in climate. Rainfall is expected to decrease by 15% by 2050. Recent modelling efforts (Farinosi & Hughes, 2020) on climate change and water-use scenarios showed that the relative impacts can be quite different across the whole ZRB, the greatest impacts being in the Lake Malawi/Nyasa sub-system, as well as other areas containing large open water bodies (natural and man-made), that are very sensitive to the combined effects of increased aridity. In addition, rainfall will be characterized by delayed onset with shorter and more intense rainfall events. This will have a negative impact on annual streamflow. This will ultimately affect agriculture, municipal, hydropower and ecosystems services at large (Beilfuss & Nhemachena, 2017).

CS11.3.2 Groundwater

The ZRB average annual groundwater recharge is estimated at 130 km^3. The International Groundwater Resources Assessment Centre (IGRAC) reports 10 transboundary aquifers; four of which are located inside the ZRB perimeter and six which are partly located within the ZRB (Cai *et al.*, 2017). Agricultural activities are the primary use of groundwater within the riparian states. Other ground water uses extend to fisheries and livestock watering. Ground water available for irrigation is estimated to be 38.5 km^3 (Altchenko & Villholt, 2015) (Figure CS11.2) whilst the irrigation potential is an estimated 2.55 million ha.

CS11.3.3 Agriculture

Land use in the ZRB is characterized by irrigated agriculture and rainfed agriculture. The WorldBank (2010) estimated that 70% of the riparian inhabitants are subsistence farmers. Stark rainfall variations are observed amongst ZRB member states with the lower parts of the basin receiving approximately 500 mm in the extreme south and southwestern parts, whereas the upper sub-basins such as Kabompo, Upper Zambezi, Lungue Bungo, Kafue, Shire, and Zambezi Delta receiving an estimated 1400 mm. The variations in rainfall consequently influence agricultural productivity, that is, northern parts experience high yield as compared to the southern regions. Irrigated agricultural practices in the ZRB are informal irrigation by small-scale farmers, smallholder irrigation, and commercial irrigation schemes. Informal irrigators have an average landholding of 200 m^2 and they make use of conventional and traditional methods such as watering cans, bucket systems, and hosepipes (Manzungu *et al.*, 2017) whereas commercial irrigation is done on a large scale and it is characterized by advanced technology and heavy machinery. Riparian governments are heavily invested in smallholder irrigation schemes (SIS) to alleviate poverty. The SIS are characterized by a landholding of 1 ha and the farmers share common pool resources (Dirwai *et al.*, 2019).

Fishing among the riparian inhabitants is done to either augment dietary needs or for small-scale commercial (income generation). Fishing in the ZRB is also done on a commercial scale. The activity is also done for angling tourism. In 2016, African Union – Interafrican Bureau for Animal Resources (AU-IBAR) reported $\approx$100 fish species within the Zambezi River with the upper Zambezi boasting more species ($>$85). Despite the importance of fishing in the riparian countries, there are still several reports of underutilized potential of fishing within the river basin and no increases in catch yields over the past decade (Table CS11.1) (AU-IBAR, 2016; Tweddle & Peel, 2015; Tweddle & Tweddle, 2010). The riparian countries have the potential to increase their catch yields, this will subsequently improve the livelihoods of the riparian population.

Livestock farming is also an important aspect in the ZRB. According to SADR (2015), the first livestock to enter Southern Africa was through the ZRB. Livestock farming consumes $\approx$120 million m^3 per annum, which is less than 1% of total consumptive use (Euroconsult Mott McDonald, 2007; World Bank, 2008). Cattle production dominates the livestock farming landscape. Cattle population within the basin is $\approx$ 42 million heads (ZAMCOM, SADC, SARDC, 2015). The number is still considered below potential. Small-scale livestock production depends on natural grasslands, whilst at the commercial production

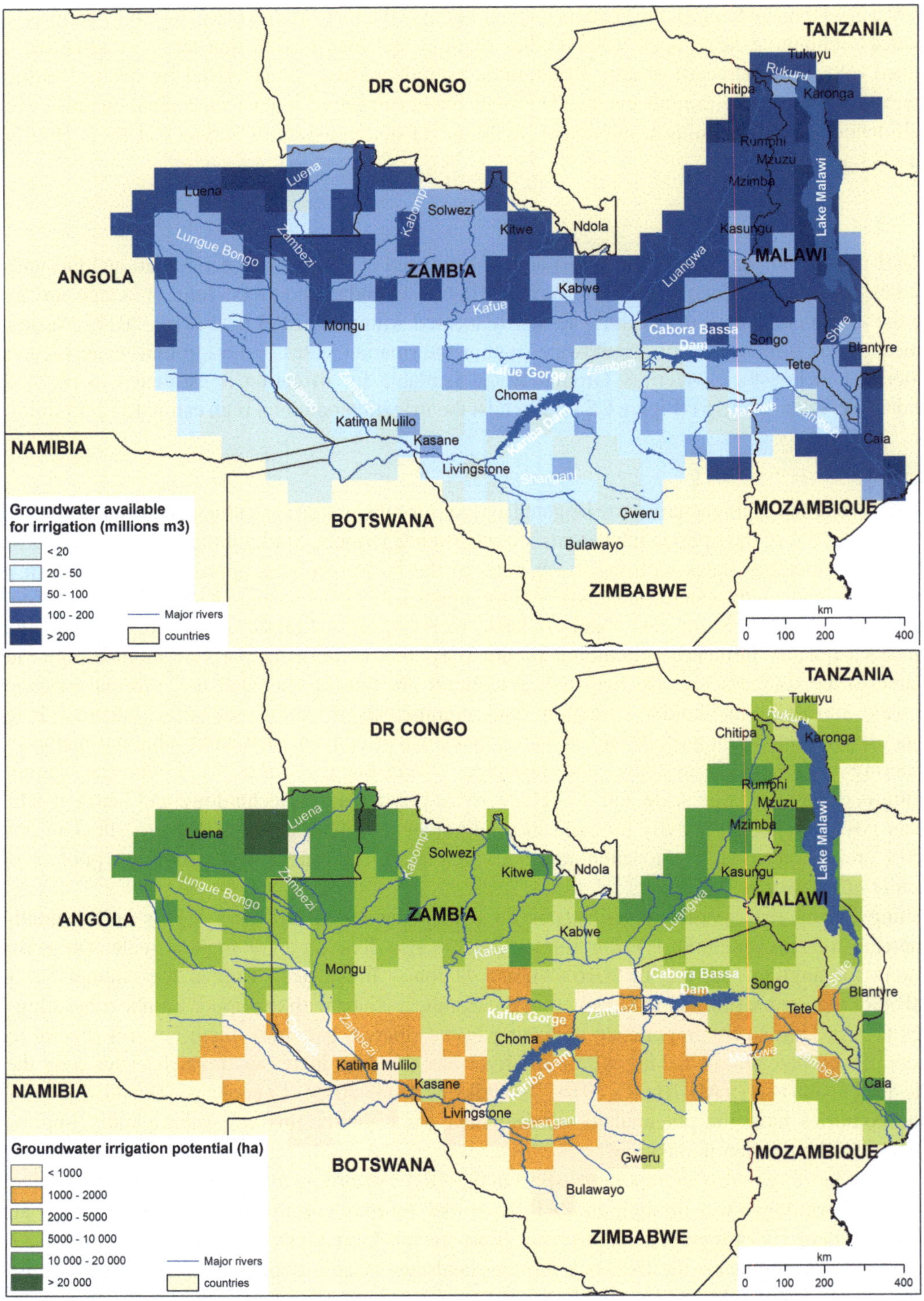

Figure CS11.2 Ground water available for irrigation.

Table CS11.1 Annual catch yields in the different regions of the ZRB in 2000–2007.

Region	2000	2001	2003	2004	2005	2006	2007
				Tonnes			
Upper Zambezi	6728	—	6694	6834	6653	6079	7421
Kariba	8863	9306	8818	9003	8768	8008	9776
Kafue	6131	6437	6100	6228	6062	5539	6763
Lukanga	1306	1371	1299	1327	1291	1180	1441
Itezhi-tezhi	2221	2332	2210	2256	2196	2007	2450
Lusiwashi	2139	2246	2128	2173	2115	1933	2359
Lower Zambezi	588	617	585	597	581	531	649
Total	29,976	24,374	29,837	30,422	29,671	27,283	32,866

Adapted from Tweddle & Tweddle (2010).

level, the herd is given supplementary feed. Erratic rainfall patterns and droughts impose production penalties on farmers that rely on grasslands.

CS11.4 WATER–ENERGY–FOOD–ECOSYSTEM NEXUS

The WEFE nexus describes the close interlinkages of the water, energy, and food sectors, and how they rely on and impact ecosystems. Primarily it focuses on the interdependencies between achieving water, energy, and food security for human well-being, that is, basic services and economic development, while ensuring ecologically sustainable use of globally essential resources. In practical terms, the WEFE Nexus helps to improve understanding and systematic analysis of the interactions between the natural environment and human activities in these three sectors. The pressure on natural resources (water, land, and energy) and the need for harmonious development while sharing transboundary resources in the ZRB demand holistic approaches to the management of such resources. The WEF Nexus is best placed as a tool for such resources' sustainable management. An exploratory WEFE Nexus analysis for the ZRB was undertaken. The main reason for the WEFE Nexus analysis are two-fold; first it allows for the inclusion of ecosystems goods and services (EGS) to the original WEF Nexus. EGS are very important in the ZRB as water and other natural resources serve many purposes for the benefit of all life in the basin (more is discussed about this in the sections to follow). Second, the WEFE Nexus, as discussed above, allows for the analysis of the water, energy, and food sectors, and their reliance on the ecosystems and the consequent impact on the same ecosystems. The WEFE Nexus indicators for the ZRB are presented in Table CS11.2 for the 2018 base year, based on latest available data.

The WEFE Nexus analysis for the ZRB proved challenging to undertake because of data issues – both spatially and temporally, as well as conflicting data from different sources. Be that as it may, the following is deduced from the analysis. On average the available water resources per capita in the basin are on the lower side at (503.25 m^3/capita) and considered unsustainable since they fall below 1700 m^3/capita. The irrigated crop water productivity, similarly is on the lower side falling below US\$10/m^3 and thus considered unsustainable. This is not surprising given the low levels of production by the smallholder irrigation projects in the basin typified by production for own consumption and some sale. With respect to energy, access to electricity was generally referred to as being low for the bulk of the basin population although

Table CS11.2 WEFE Nexus indicators for the ZRB.

WEFE Nexus	WEFE Nexus Indicator	Status	Notes
Water	Proportion of available freshwater resources per capita (availability)	503.25 m^3/capita	(1)
	Proportion of crops produced per unit of water used – irrigated (productivity)	US\$2.01/m^3	(2)
Energy	Proportion of the population with access to electricity (accessibility)	Very low	(3)
	Energy intensity measured in terms of primary energy and GDP (productivity)	0.23×10^{-6} MW/GDP 0.46×10^{-6} MW/GDP	(4)
Food	Prevalence of moderate or severe food insecurity in the population (self-sufficiency)	No data at ZRB level	(5)
	Proportion of sustainable agricultural production per unit area (cereal productivity)	1.16 tons/ha (maize) 1.13 tons/ ha (paddy rice)	(6)
Ecosystem	Proportion of ecosystems goods and service value per capita (value)	US\$36.05/capita	(7)
	Water provisioning in ecosystems goods and service per capita	473 100 m^3/capita 25 m^3/capita (excluding evaporation)	(8)
	Environmental flow requirements (sustainability)	1.16% MAR	(9)

Notes: (1) From ZRB book Chapter 2 & EuroConsult Mott MacDonald (2007 – Table 2.4); (2) from ZRB book Chapters 1 and 6 in "ZRB – Water and Sustainable Development". In other reports the Agricultural GDP in the ZRB is given as US\$14 billion (World Bank, 2010 – Table 3.57); (3) no data; (4) from ZRB book Chapter 5 and Table 5.1 Tilmant (2017); (5) mixed data and none at ZRB level; (6) from World Bank (2010 – Table 3.69); (7) from ZRB book Chapter 7 and McCartney and Nyambe (2017); (8) from ZAMCOM *et al.* (2020) Table 7.4; (9) from ZRB book Chapter 2 and EuroConsult Mott MacDonald (2007) Table 2.4.

the ZRB is considered one of the energy generation hubs of the region. Cereal productivity is considered marginally sustainable at 1.12 tons/ha for maize and 1.13 tons/ha for rice – the two cereals that constitute the base of food for the rural population. Again, these levels of productivity are not surprising given the dominance of small-scale agricultural production based on small areas and low input of agro-chemicals and the lack of access to credit facilities for intensified production. By extension, such low levels of production coupled with high population increases in the region inevitably lead to food insecurity problems in the basin.

In terms of the economic value of EGS, these worked out at US\$36.05 per capita, which when converted to a base of a day (so as to compare this to the concept of \$/capita/day), comes out low. Although this measure is considered low, it is acknowledged that a large proportion of the basin population derives its livelihood from these EGS. In terms of water provisioning as an EGS, this seems healthy at 473,100 m^3/capita but drops to 25 m^3/capita when evaporation is excluded, translating to about 70 l/capita/day. In terms of environmental flows, data from the ZRB is conflicting. Data base of river flows seem to indicate environmental flows of 15–20% of the MAR, but when this is based on EGS, the environmental flows work out at 1.14% of the MAR. Considering the lower figure, sustainability of environmental flows is poor, but reality on the ground is something else.

CS11.5 CONCLUSION AND RECOMMENDATIONS

Agriculture is the largest water consumer in the ZRB. Irrigation is the key driver of the agricultural-based economies of the basin countries with agricultural activities being dominant in Malawi, Mozambique, Tanzania, Zambia, and Zimbabwe. Apart from cropping agriculture, the other agricultural activities practices in the ZRB include fisheries and livestock farming. With respect to fisheries, catch yields within the upper Zambezi are approximately 7500 tonnes per annum and this approximately half of the potential annual yield (14,000 tonnes per annum). Livestock production only consumes about 120 million m^3 per annum (representing less than 1% of total consumptive use). Ninety per cent of the livestock in the basin is in Zambia, Zimbabwe, and Malawi. Cattle population within the basin is approximately 42 million heads having risen from 35 million in 2005. The WEFE Nexus analysis yielded some interesting results, but on the balance of issues, all indicators are marginally sustainable. This information can be used to plan for the future in a sustainable manner in the ZRB.

CS11.6 STUDY LIMITATIONS AND THE WAY FORWARD

The transboundary ZRB, the fourth largest in Africa poses many challenges from the perspective of WEFE Nexus. The report explored water and agriculture aspects in the ZRB focusing on irrigated and rainfed agriculture through appropriate agricultural water management practices. The report was more inclined to discuss crop production in the ZRB as compared to other consumptive water uses because crop production in the ZRB is the biggest water consumer in the basin. The report highlights potential areas (fisheries and livestock production) where riparian member states can focus on for improved income generation for the riparian inhabitants. A more thorough and detailed WEFE Nexus analysis is required with robust data sets to aid in the sustainable natural resources planning for the ZRB into the future.

REFERENCES

Altchenko Y. and Villholt K. G. (2015). Mapping irrigation potential from renewable GW in Africa – a quantitative hydrological approach. *Hydrology Earth Systems Science*, **19**, 1055–1067.

AU-IBAR. (2016). Interafrican Bureau for Animal Resources. African Union, Addis Ababa, Ethiopia.

Beilfuss R. D. and Nhemachena C. (2017). Climate change vulnerability and risk. In: The Zambezi River Basin, J. Lautze, Z. Phiri, V. Smakhtin and D. Saruchera (eds.), Routledge, Oxfordshire, United Kingdom.

Cai X., Altchenko Y. and Chavula G. (2017). Availability and use of water resources. In: The Zambezi River Basiin Water and Sustainable Development, J. Lautze, Z. Phiri, S. Vladimir and D. Saruchera (eds.), Routledge, London, United Kingdom, pp. 7–28.

Dirwai T., Senzanje A. and Mudhara M. (2019). Water governance impacts on water adequacy in smallholder irrigation schemes in KwaZulu-Natal province, South Africa. *Water Policy*, **21**(1), 127–146.

Euroconsult Mott McDonald (2007). Rapid Assessment – Final Report Integrated Water Resources Management Strategy for the Zambezi River Basin. Mott MacDonald Limited Group, London, United Kingdom.

Farinosi F. and Hughes D. (2020). Assessing development and climate variability impacts on water resources in the data scarce Zambezi River basin. Part 2: Simulating future scenarios of climate and development. (Manuscript Draft under review by *Journal of Hydrology: Regional Studies*).

Gomo F. F., Macleod C., Rowan J., Yeluripati J. and Topp K. (2018). Supporting better decisions across the nexus of water, energy and food through earth observation data: case of the Zambezi Hulsmann, S., Susnik, J., Rinke, R., Langan, S., van Wijk, D., Janssen, ABG., Mooij, WM. 2019. Integrated modelling and management of water resources: the ecosystem perspective on the nexus approach. Current Opinion in Environmental Sustainability 2019, Vol 40:14–20 https://doi.org/10.1016/j.cosust.2019.07.003 basin. *Proceedings of the International Association of Hydrological Sciences*, **376**, 15–23.

Kling H., Stanzel P. and Preishuber M. (2014). Impact modelling of water resources development and climate scenarios on Zambezi River discharge. *Journal of Hydrology: Regional Studies*, **1**, 17–43.

Manzungu E., Senzanje A. and Mutiro J. (2017). Towards sustainable agricultural water management. In: J. Lautze, Z. Phiri, V. Smakhtin and D. Saruchera, The Zambezi River Basin. Routledge, Oxfordshire, United Kingdom.

McCartney M. and Nyambe I. A. (2017). Ecosystem services: opportunities and threats. In: The Zambezi River Basin, J. Lautze, Z. Phiri, V. Smakhtin and D. Saruchera (eds.), Routledge, Oxfordshire, United Kingdom.

Pinay G. (1988). Hydrobiological assessment of the Zambezi river system: a review.

Sadr K. (2015). Livestock first reached southern Africa in two separate events. *PloS One*, **10**(8): e0134215. https://doi.org/10.1371/journal.pone.0134215

Schleiss A. and Matos J. (2010). Zambezi River Basin. pp. 1–8.

Tilmant A. (2017). Hydropower and the water-energy-food nexus. In: The Zambezi River Basin, J. Lautze, Z. Phiri, V. Smakhtin and D. Saruchera (eds.), Routledge, Oxfordshire, United Kingdom.

Tweddle D. and Peel R. A. (2015). Challenges in fisheries management in the Zambezi, one of the great rivers of Africa 99–111. doi: 10.1111/fme.12107

Tweddle D. and Tweddle D. (2010). Aquatic ecosystem health & management overview of the Zambezi river system: its history, fish fauna, fisheries, and conservation. *Aquatic Ecosystem Health & Management*, **4988**, 224–240. doi: 10.1080/14634988.2010.507035

World Bank. (2008). Zambezi River Basin Sustainable Agriculture Water Development. World Bank, Washington, DC, USA.

WorldBank. (2010). The Zambezi River Basin: A Multi-Sector Investment Opportunities Analysis. World Bank, Washington DC, USA.

ZAMCOM. (2020). Livestock and Aquaculture Census Report. Ministry of Fisheries and Livestock Central Statistical Office, Lusaka, Zambia.

ZAMCOM, SADC, SARDC. (2015). Zambezi Environment Outlook. Harare, Gaborone. Botswana.

Zimba H., Kawawa B., Chabala A., Phiri W., Selsam P., Meinhardt M. and Nyambe I. (2018). Regional Studies Assessment of trends in inundation extent in the Barotse Floodplain, upper Zambezi River Basin: a remote sensing-based approach. *Journal of Hydrology: Regional Studies*, **15**, 149–170. doi: 10.1016/j.ejrh.2018.01.002

Renewable energy for irrigation in the MENA Region

R. H. Mohtar[1,2]

[1]*Texas A&M University, USA*
[2]*American University of Beirut, Lebanon*

ABSTRACT

The SDGs bring new momentum, allowing renewable energy to be a catalyst of equity, alleviation of poverty, and access to primary resources (water, energy, food) in remote areas that were without access and opportunities for these primary resources in the past. This contribution to the position paper presents a framework to increase renewable energy penetration in the Arab countries of the MENA region in support of irrigated agriculture. It presents technological, policy, capital, and human capacity challenges and opportunities for renewable energy sources, including solar, wind, hydropower, and bioenergy. The contribution concludes by discussing readiness for renewable energy in the Arab world, and is intended to help guide discussions towards the renewable energy transition. The core and cross-cutting role of water in achieving multiple SDGs can be observed with clarity: water availability and access is directly linked with poverty, health, economic growth, education, social justice, as well as food and energy securities. It has been reported that blackouts due to water shortages in some regions on Africa can cause an annual decrease of 2–4% in GDP.

Keywords: Renewable energy, irrigation, MENA region

CS12.1 INTRODUCTION

CS12.1.1 Irrigation demands and gaps

Rain-fed agriculture accounts for two thirds of the total cropland in the Middle East and North Africa (MENA). This region is also one of the most water scarce in the world, having an annual per capita water availability of 1100 m^3 (compare to the global average of 8900 m^3 per year). By 2025, this is projected to reach only 550 m^3 per year (World Bank, 2012).

Climate change models project several scenarios for the MENA. Most climate models for MENA project drier conditions; all projections conclude that the demand for irrigation water will increase between 15 and 33% by 2050 (Figure CS12.1). In addition, as the climate results in drier conditions, available water

	2000–09	2020–30	2040–50
Total Demand	261	319	393
Irrigation	213	237	265
Urban	28	50	88
Industry	20	32	40
Total Supply	219	200	194
Surface water[a]	171	153	153
Groundwater	48	47	41
Total Unmet demand	42[b]	119	199
Irrigation	36	91	136
Urban	4	16	43
Industry	3	12	20

Source: FutureWater 2011.
a. Surface water includes river flows into the MENA Region.
b. Summation does not add up due to rounding.

Figure CS12.1 MENA annual water demand and supply (km^3) under average climate change scenario (between 2000 and 2050).

resources in the region could be reduced by more than 40%.[4] For these reasons and because climate change can also decrease crop yields and affect water productivity, the water gap in the agriculture sector is a major concern for the water, energy, and food securities of the region. Many countries have counteracted this water gap by over-extracting groundwater. Hydrological models predict that the groundwater recharge will decrease up to 40% in the Gulf region alone: even the wetter parts of MENA are expected to experience a considerable reduction (38% in Morocco, 34% in Iraq, 22% in Iran).[5]

Supplemental irrigation in the region is a must for counteracting the effects of climate change and potentially increasing yields in MENA's different agro-ecological zones. There is a gap between the actual and potential yields with supplemental irrigation (between 0.5 and 5 t/ha), therefore, there is vast room for improvement (Anderson *et al.*, 2016). A wide range of strategies and technologies are currently available that could potentially increase yields, but these approaches must be site specific.

For example, ICARDA has shown (Figure CS12.2) in recent studies that small amounts of supplemental irrigation in rain-fed croplands can increase wheat grain yields by 30% up to 400%, depending on the rainfall at a specific region in Syria (Haddad *et al.*, 2011).

On the other hand, irrigated agriculture is the largest consumer of ground, and freshwater more generally, in the region, and the challenges are real, including: (1) securing sufficient water supply; (2) increasing water productivity (amount of water required per ton of produce); (3) providing reliable energy to transport water and make it available in all croplands; and (4) reducing the energy footprint of these operations. Synergies and trade-offs in the WEF Nexus are evident and are affected by external factors such as geographical conditions, socio-economic aspects, local human capacity, and policies.

---

[4]Idem.

[5]Idem.

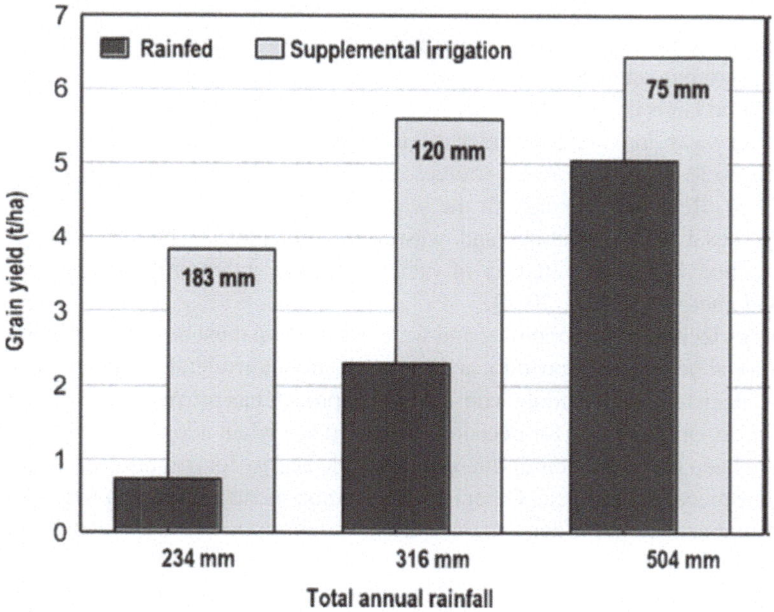

Figure CS12.2 Impact of supplemental irrigation on rainfed wheat yield in northern Syria in dry, normal, and wet years, with supplemental irrigation of 183, 120, and 75 mm. *Source*: Oweis and Hachum (2009).

CS12.2 SOIL–WATER–PLANT RELATIONSHIP AND THE NEED FOR NEW AGRICULTURE MODEL

A quantitative analysis of the agricultural water gaps is important moving forward; however, the food production system community must also look afresh towards securing a sustainable agriculture system. In addition to crop yield and food production, such a system should also consider the nutritional aspects of food production and the values of other resources used in food production. These resources include the water and energy footprints, land and soil quality, air and carbon emission. With such systems level consideration, a WEF Nexus trade-off analysis emerges that dictates certain optimal levels of a food production system that is renewable and sustainable from a holistic perspective and at the economic, social, and environmental levels (Mohtar, 2017; Mohtar *et al.*, 2017).

CS12.3 RENEWABLE ENERGY, THE WEF NEXUS AND THE SDGS

The interconnectedness of the WEF Nexus makes it clear that the energy demands of agricultural processes will have a significant impact on food and water securities. Remote areas that are not connected to the grid may be the most vulnerable in terms of securing energy requirements for food and water production. Therefore, the technologies that countries include in their energy portfolio will not only impact SDG 7 (ensuring access to affordable, reliable, sustainable, and modern energy for all), but will also affect the capacity of a country to ensure the availability of water and sanitation (SDG 6) and the achievement of food and nutrition security through the promotion of sustainable agriculture (SDG 2). Decision support tools that use a nexus approach to evaluate the effects of different energy mixes is highly valuable to ensuring food and water securities, not only at local but also at a regional (transboundary) level.

Coherent policy making and strategic planning should go beyond simple synergies and trade-offs to take into account the multiple interlinkages between the sectors.

Water transport (ground water and surface pumping) represents a major energy requirement in irrigated croplands. Agricultural lands that are physically far from available fresh water face considerable high energy costs for transporting water, making the issue of affordability very important. Conversely, some croplands in remote areas, even if closer to fresh water sources, might not be connected to the grid and therefore face access issues as well. Renewable energy in the water supply industry can play a key role in operations such as pumping, desalination, heating, and wastewater treatment. Although energy from renewable sources will not reduce the energy intensity of such operations, it certainly will be crucial for enhancing access to water and energy (IRENA, 2014).

Renewable energy technologies for power and water production must be carefully evaluated before being adopted into a national or regional portfolio, and must take into consideration the broader impacts on other sectors. The use of decision support tools with a Nexus approach has proven to provide relevant qualitative and quantitative information crucial for decision makers to see when adopting new technologies.

It is important to keep in mind that introducing renewable energy for irrigation and other food production operations is but one piece of the puzzle. Other important components for moving into renewable agriculture include crop improvement through genetics, cropping system optimization, water harvesting to increase water availability to crops, and other practices to ensure the sustainability of food production at economic, social, and environmental levels. Water harvesting practices are not new to the region: some of the oldest practices exist there. These can be localized to store water directly in the root zone, in cisterns or reservoirs, and be integrated with irrigation practices. Therefore, a holistic water–energy–food system is necessary to augment technological solutions to food production.

CS12.4 RENEWABLE ENERGY IN THE ARAB WORLD

Figure CS12.3 shows the availability of renewable energy sources in the Middle East and North Africa region. This map agrees with data interpolated from a global dataset produced by NASA's Surface meteorology and Solar Energy (SSE) program. It is clear that the Arab region shows significant resource availability for solar energy (NASA, 2014). The Atlantic part of North Africa has a significant potential for wind energy.

It is interesting to observe how renewable energy has integrated into existing energy systems, having a complementary, non-'competing' role along with conventional power. This is important especially for countries looking to diversify their energy mix and economy and for which the target is to break dependency on a single fuel.

The choices of renewable energy (RE) technologies in the water sector are highly dependent on local geographical and climate conditions as well as human capacities and cost. Developing low-cost local technologies remains a big challenge throughout the MENA region.

Renewable energy source water pumping systems can be described in five major groups: (1) solar photovoltaic systems, (2) solar thermal systems, (3) wind energy systems, (4) bioenergy systems, and (5) hybrid renewable energy water pumping systems (Gopal *et al.*, 2013). For the MENA, solar photovoltaics is the most widely used technology, and the one that most makes sense, given the climatic and geographical conditions. Solar photovoltaics are followed by wind pumping systems. Solar thermal and bioenergy systems are unpopular. Hybrid solar/wind systems would make sense in the region and have a lot of potential, but the technology is not fully developed.

Success stories and case studies using RE technologies for water pumping are widely reported. These technologies still need further research for wider implementation include: optimization of PV

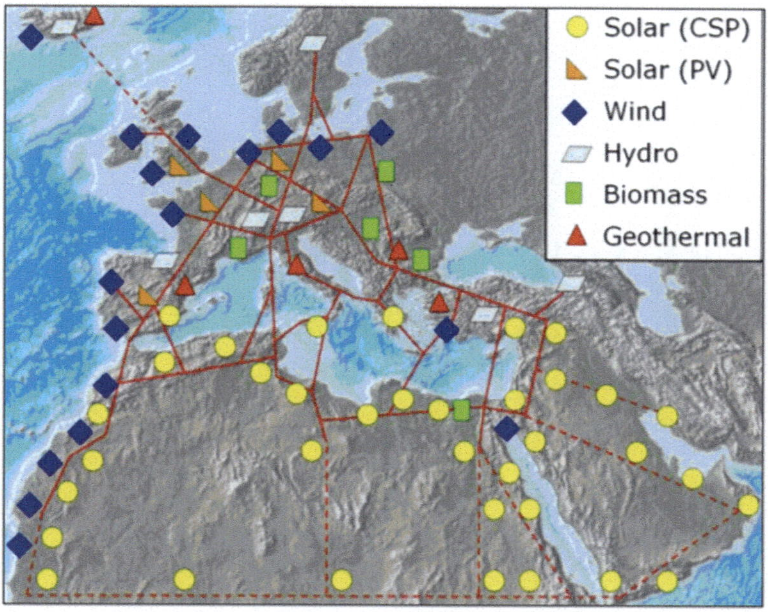

Figure CS12.3 Renewable energy map for the Arab world. *Source*: DESERTEC (2013).

panels (tilt angles), cooling of PV panels, reducing dust accumulation in PV panels, development of new materials for PV, and reduction of power loss, among others.[6]

Technology choices should be evaluated systematically, taking into consideration the six main factors described in the ESCWA Regional Policy Toolkit (ESCWA, 2016):

(1) Resource requirements (sustainability – water and energy footprint)
(2) Economic aspects (all costs associated, land requirements, etc.)
(3) Environmental impact (air, water, soil, biodiversity impacts)
(4) Human capacity requirements (local skills required)
(5) Technical requirements and robustness (considering local components)
(6) Social–cultural criteria (awareness of institutional requirements, responsibilities).

CS12.5 RENEWABLE ENERGY READINESS OF A COUNTRY AND GLOBAL COMPETITIVENESS INDEX

The European Union-Gulf Cooperation Council (EU-GCC) network describes important criteria that can be used to evaluate how ready a country is to implement renewable energy technologies. Three factors are of utmost importance:

(1) **Infrastructure**: including natural resources. Country overall infrastructure, grid capacity, market infrastructure, electricity access rate, and projected demand. As mentioned, Arab agriculture faces limited natural resources.

[6]Idem.

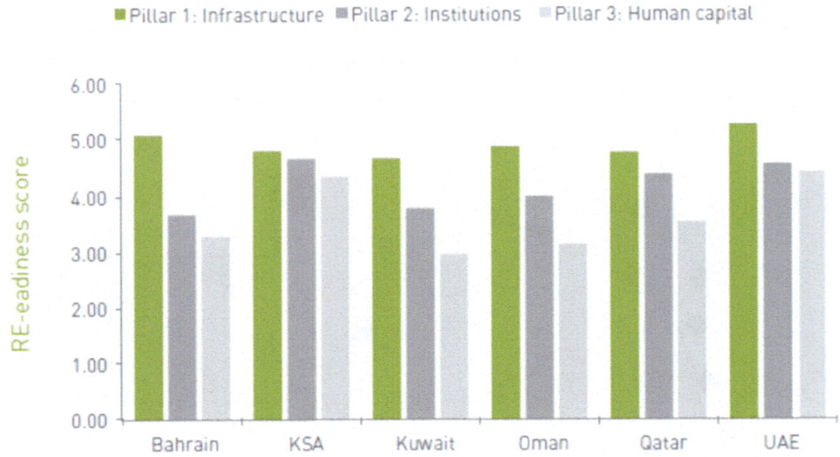

Figure CS12.4 The GCC countries' attractiveness index on renewable energy development pillars. *Source*: EU-GCC (2013).

(2) **Institutions**: both public and private institutions. Energy, key policies, access to renewable energy finance, and macroeconomic environment. Specifically in the MENA, budgets for R&D are insufficient to localize technologies, adopt them, and implement policies.

(3) **Human capital**: technical and commercial skills, technology adoption and diffusion and awareness among consumers, investors, and decision makers. For example, in the Arab world, there is little work being done towards crop improvement: technologies are available but their adoption is hindered by geographical, economic, and human capacity realities.

Taking into account these three main criteria or pillars, the EU-GCC study generated scores for several countries in the Gulf sub-region. The methodology defined in this study could be used to generate similar data for other countries in the MENA. The scores shown indicate that, in the GCC countries, the human capital factor for renewable energy deployment is the weakest factor (Figure CS12.4); the policy and institutional aspect comes in the middle; and the infrastructure scores indicate that this aspect is relatively strong (ESCWA, 2016).

While the RE technologies have reached a relative level of maturity in global markets, these technologies are still not competitive in the MENA region. It is here that the active involvement of governments will determine how ready they will be to adopt RE into their energy mix and enter the markets at a more competitive level. It is commendable that many countries in the Region have already established national targets for their RE mix, with dates for achieving it; additional support is now required to implement policies, capacity building, localization, and deployment of technologies. A few barriers that further limit competitiveness in the region are: (1) bureaucracy and inefficient institutional structures, (2) lack of policy support, and (3) fossil fuel/electricity subsidies (ESCWA, 2016).

CS12.6 KEY FINDINGS

Distributed renewable energy has a big role to play in the water and food security of the MENA region. It can help provide the enabling energy source for distributed water and food production in remote rural areas, where access to water and food are at risk. Towards energy as an enabler to human water and food security, we note the following findings:

- A new business model for renewable and sustainable agriculture must be established for the arid and semiarid regions of MENA; this can be accomplished by looking into nutrition, water, energy, land values in addition to the economics of production.
- Technology scorecards taking into consideration the aspects mentioned above (resource requirements, economic aspects, environmental impact, human capacity requirements, technical requirements and robustness, and social–cultural criteria) can help with the choice of a technology that is suited to the local conditions.
- A renewable energy plan is needed to help transition towards a renewable energy portfolio in the MENA region; such a plan should include assessment of the regional integration of primary resources security.

CS12.7 DISCUSSION, CONCLUSIONS, AND RECOMMENDATIONS

The following concluding remarks are highlighted:

- Research is needed to understand how water scarcity affects communities at different levels and how this affects SDG implementation.
- Local data are needed to define the level of sustainable production to be achieved for each eco-zone in the MENA; these eco-zones are areas with similar social, environmental, and economic conditions, and wherein the elements of the new model for agriculture described above can be established.
- Implementation of a set of guidelines establishing which technology is most suited for local conditions in the specific region of MENA.
- Last but not least, a regional effort to develop a renewable readiness plan must be established to help countries in the region achieve their goals and to share knowledge in implementation of renewable and nonrenewable energy integration. These plans should focus on various elements including, policies and incentives, localization of knowledge, capacity building, and manufacturing and industries.

REFERENCES

Anderson W., Johansen C. and Siddique K. M. (2016). Addressing the yield gap in rainfed crops: a review. *Agronomy for Sustainable Development*, **36**, 18.

DESERTEC (2013). Global 'Energiewende': The DESERTEC concept for Climate Protection and Development. Available from http://www.desertec.org/#!desertec-atlas/pznkq.

ESCWA (2016). Developing the Capacity of ESCWA Member Countries to Address the Water and Energy Nexus for Achieving Sustainable Development Goals: Regional Policy Toolkit. ESCWA, Beirut.

EU-GCC (2013). Renewable Energy Readiness Assessment Report: The GCC Countries 2011–2012. Masdar Institute of Technology. Study under EU-GCC Clean Energy Network 2013. Available from http://eugcc.epu.ntua.gr/LinkClick.aspx?fileticket=ZLVAyWw9x_8%3d&tabid=448.

Gopal C., Mohanraj M., Chandramohan P. and Chandrasekar P. (2013). Renewable energy source water pumping systems – a literature review. *Renewable and Sustainable Energy Reviews*, **25**, 351–370.

Haddad N., Duwayri M., Oweis T., Bishaw Z., Rischkowsky B., Aw-Hassan A. A. and Grando S. (2011). The potential of small-scale rainfed agriculture to strengthen food security in Arab countries. *Food Security*, **3**(Suppl 1), 163. https://doi.org/10.1007/s12571-010-0099-7.

International Renewable Energy Agency (IRENA) (2014). Pan Arab Renewable Energy Strategy 2030. Available online at http://www.irena.org/DocumentDownloads/Publications/IRENA_Pan-Arab_Strategy_June%202014.pdf.

Mohtar R.H. (2017). A call for a new business model valuing water use and production: the water, energy and food nexus holistic system approach. *Water International*, **42**(6), 773–776. Groundwater and Climate Change – Multi-Level Law and Policy Perspectives. doi: 10.1080/02508060.2017.1353238.

Mohtar R. H., Assi A. T. and Daher B. T. (2017). Current water for food situational analysis in the Arab Region and expected changes due to dynamic externalities. In: The Water, Energy, and Food Security Nexus in the Arab Region, K. Amer, Z. Adeel, B. Böer and W. Saleh (eds.), Cham: Springer International Publishing, pp. 193–208. https://doi.org/10.1007/978-3-319-48408-2_10

National Aeronautics and Space Administration (NASA) (2014). Surface meteorology and Solar Energy: A renewable energy resource web site (release 6.0). Available from http://eosweb.larc.nasa.gov/sse/.

Oweis T. and Hachum A. (2009). Supplemental irrigation for improved rainfed agriculture in WANA region. In: Rainfed Agriculture: Unlocking the Potential, S. P. Wani, J. Rockström and T. Oweis (eds.), Comprehensive Assessment of Water Management in Agriculture Series 7. CABI, London.

World Bank (2012). Renewable Energy Desalination: An Emerging Solution to Close the Water Gap in the Middle East and North Africa. (MENA development report). ISBN 978-0-8213-9457-1.

The water–energy–food Nexus in the Arab region: governance and role of institutions

W. K. Al-Zubari

Water Resources Management Program, Department of Environmental and Natural Resources, College of Graduate Studies, Arabian Gulf University

ABSTRACT

Interlinkages between water, energy, and food systems in the Arab region are intensifying with time, as the demand for resources increases due to population growth and is exacerbated by changing consumption patterns, low management efficiencies, and the additional impacts of climate change. While these conditions are constraining the ability of existing systems to meet the growing demand in a reliable and affordable manner, the dynamics of the three sectors pose substantial risks for the sustainable development and resource security ambitions of the Arab countries due to their strong and crucial interdependencies. Therefore, it is becoming imperative that policy formulation becomes coordinated among the three sectors. Conventional policy and decision making in 'silos' needs to give way to a nexus approach to reduce trade-offs and build synergies across the three sectors through integrated planning and management, which can only be delivered through appropriate governance and relevant institutions. Currently, the institutional framework governing the elements of the nexus in most of the Arab countries is fragmented and lacks effective coordination mechanisms, which has led to a sectorial approach to policy planning, and consequently fragmented strategies and policies. Some countries have succeeded in presenting different models of 'integrated institutions,' but their comprehensive and inclusive management of these interlinked priorities still needs support. Enhancing coordination and collaboration mechanisms amongst institutions is key for mainstreaming the WEF Nexus approach and not necessarily establishing new institutions.

Keywords: WEFE Nexus, Arab region, Governance

CS13.1 INTRODUCTION

Water, food, and energy in the Arab region are strongly and closely interlinked, probably more than in any region of the world. Generally, the region is energy intensive, water scarce, food deficient, and one of the world's most economically and environmentally vulnerable regions to climate change. To make enough food to support a growing population and urbanization, more water and energy are needed; to make water accessible and clean for human consumption demands energy; and producing energy will require water. These interdependencies, termed the 'Water–Energy–Food Nexus', are intensifying in the region as demand for water, energy, and food increases with population growth, changing consumption patterns, and low management efficiencies in both the supply and demand of these three sectors, and are expected to be further compounded by the impacts of climate change.

These strong interlinkages carry high risks among the three sectors, and the conditions of dwindling natural resources create immense challenges to the countries in this arid region in their attempt to meet the demands of the three sectors. Trying to achieve the security of one of these sectors independently and without due consideration of the trade-offs with the other two sectors will be at the expense of the security of one or two of the components of the nexus, and eventually endangering the security of the sector itself. For example, achieving food security by domestic production without due consideration of the limitations of water resources will not only lead to over-exploitation, quality deterioration, and loss of water resources, but it will also lead eventually to the loss of agricultural productivity, and the deterioration of the agriculture sector itself and the levels of food security.

In other words, adapting a sectorial approach in meeting the demand of these sectors will lead to worsening livelihoods and increasing environmental degradation and thus, potentially, missed opportunities to achieve the SDGs and ultimately negative impacts on human wellbeing. Hence, such a strong interdependency between these three sectors and between them and the environment and climate change calls for a nexus thinking and perspective when addressing the planning and management of these three vital sectors (i.e., an optimization approach rather than maximization); this is an approach that integrates management and governance across sectors, and where conventional policy and decision-making in 'silos' gives way to an approach that reduces trade-offs and builds intelligent synergies across the sectors. Adopting a WEF Nexus approach in the Arab countries would provide an opportunity for innovation and learning to minimize security risks and maximize opportunities and enhance resource efficiency and equity. More importantly, it will serve the countries of the region in moving towards achieving the global SDGs and meeting the mandates of a low carbon economy following from their committed Nationally Determined Contributions (NDCs) under the Paris 2015 Climate Change agreement.

Fortunately, this has been well recognized in the region through the 'Arab Strategic Framework for Sustainable Development (ASFSD)', adopted by the League of Arab States (LAS) in 2013, aiming to address the key challenges faced by the Arab States in achieving sustainable development during the period 2015–2030. This new development has created unprecedented opportunities for fundamental policy changes in various economic, institutional, technological, and social systems, as well as boosting resource efficiency and productivity by addressing externalities across sectors.

While the usefulness of adopting a nexus approach to achieve the post-2015 sustainable development Agenda (directly SDGs 2, 6, and 7, and indirectly many others) is evident, there are still many questions that need to be addressed on the implementation. Among the most important ones are questions related to mainstreaming the WEF Nexus into institutional and policy frameworks and required capacity development, characterization, and research on the WEF Nexus to support the implementation of the SDGs.

Harvesting synergies and reducing negative trade-offs across sectors and resources, including efforts to increase cross-resource efficiencies and subsequently to also increase water, energy, and food (and political) security, requires coordination and cooperation across institutions, strategies, policies, and activities. Only a coordinated approach will ensure that efforts in one sector do not cause harm in other sectors but complement each other and generate synergies.

This short paper presents the status and main characteristics of the WEF Nexus in the Arab region, and an overview of the existing institutional framework governing the Nexus elements in the region.

CS13.2 STATUS OF WATER, ENERGY, AND FOOD IN THE ARAB REGION

The WEF Nexus in the Arab region is being driven by many natural, demographic, and socio-economic factors that do not only intensify the nexus interlinkages in the region, but also increase the risks of the nexus components on each other. In the last three decades, most of the countries in the Arab region have experienced rapid population growth and accelerated socio-economic development; the population of the region has doubled from about 170 million inhabitants in 1980 to more than 375 million in 2014 (UNDESA, 2015). This growth is associated with a substantial increase in the demands for water, energy, and food. However, the increase is attributed not only to population growth, but also to consumption patterns and low efficiencies in the production, supply, and use of these three vital resources.

CS13.2.1 The energy sector

The energy sector is central to development in the Arab region, making up about 40% of the region's GDP (Fattouh & El Katiri, 2012). The region's fossil fuel engines of growth are driven by the group of high-income, energy-rich economies of the Arabian Gulf; with record prices in recent years, this has translated into a surge of public revenue streams and related development initiatives. With rapid growth in energy-intensive urban/industrial growth, these countries are seeing some of the world's fastest rates of per-capita energy consumption growth, which is resulting in rapid draining of oil and gas reserves of these countries. This has become a top concern in most of these countries given its criticality for future exports revenues and fiscal stability (Khody & Gitonga, 2017).

Meanwhile for the region's energy-import dependent countries, energy challenges are of different types and greater. These countries make up the majority of the Arab countries and host 40% of the region's poor who still lack adequate access to modern energy services (Koday, 2012). These energy challenges, as well as challenges of other resources, are further exacerbated due to the flow of refugees and forced migrants to these countries (e.g., Jordan and Lebanon). Another challenge facing these countries is the volatility of global energy prices, which has been a drain for public revenues and, coupled with relatively high domestic energy subsidies, is creating major challenges to fiscal sustainability (Sdralevic *et al.*, 2014).

Energy consumption in the region continues to be dominated by fossil fuels. In 2011, the primary energy consumption mix was dominated by oil products (48.5%) and natural gas (50%), with coal (0.7%) playing a minor role and hydro-electricity (0.8%) being the only form of renewable energy to make a measurable impact (AFED, 2013). Current trends and patterns of energy use put the Arab countries' economies among the least efficient ones in global comparisons. Moreover, there has been no decoupling between economic growth and energy demand in the region in the past decade. Growth in energy consumption has been faster than economic growth during the past decade (Figure CS13.1), implying energy is not being used effectively to produce value within the region economies (RECREE, 2015).

Fossil fuel subsidies are a contributing factor to the inefficient use of energy. In Arab electricity markets, price subsidies represent one of the major challenges to progress of efficiency measures. Another factor is the prevalence of inefficient electricity infrastructure in most countries of the region. Average Arab electric energy losses in generation, transmission, and distribution are 19.4%, which is higher than the world average (8.3%), and much higher than the EU average (5.8%), thus presenting an ample opportunity for achieving energy savings (AFED, 2013).

CS13.2.2 The water sector

The Arab region has an extremely poor supply of water resources with many areas experiencing unpredictable rainfall. Taking population size and growth into consideration, the Arab region is considered one of the world's most water-stressed regions, with continuously decreasing per-capita

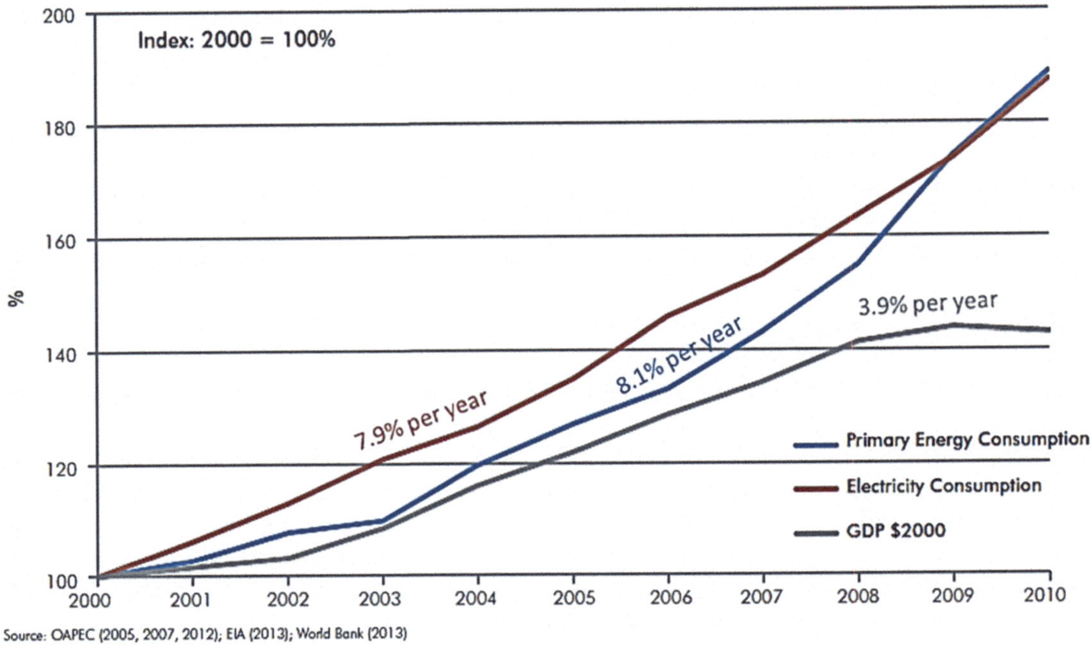

Figure CS13.1 Decoupling of energy consumption from economic growth in the Arab region, 2000–2011 (AFED, 2013).

freshwater availability. The majority of the Arab countries are currently below the water poverty line of 1000 m^3/capita/year, in contrast to a world average of about 7240 m^3/capita/year (Figure CS13.2). In 2011, the overall per capita freshwater availability in the Arab region was about 800 m^3/capita/year. Based on the projected population increase, it is expected that this indicator will continue to decrease to reach about 500 m^3/year by 2030 when the Arab region population will reach more than 500 million (Al-Zubari, 2017). This means that the whole region will experience absolute water poverty, whereby water will become a major constraint for development impacting the standard of living, health, and the environment (Falkenmark, 1989). In addition, precipitation trends are predicted to be decreasing in most of the Arab region while temperature will be increasing (ESCWA *et al.*, 2017). It is expected that by 2030 climate change will have led to a 20% reduction in renewable water resources and more droughts in the region (Doumani, 2008), which would further exacerbate the current scarcity situation.

Furthermore, one of the major challenges facing the Arab region is the high overall dependency ratio of the region on shared water resources (i.e., external water resources); more than 60% (about 174 billion m^3/year of a total of 315 billion m^3/year) of surface water resources originate from outside the Arab region (ESCWA & UNEP, 2015; Figure CS13.2). This issue remains a major concern threatening the region's stability, food security, and complicates national water resources management and planning. Conventions and agreements on equitable sharing and management of water resources have not been signed by riparian countries. In addition, some Arab countries are deprived of their water resources by occupying powers (i.e., Occupied Palestinian Territories, Golan Heights, and Southern Lebanon) which is another major issue in the region and is constraining the development of the population of these countries.

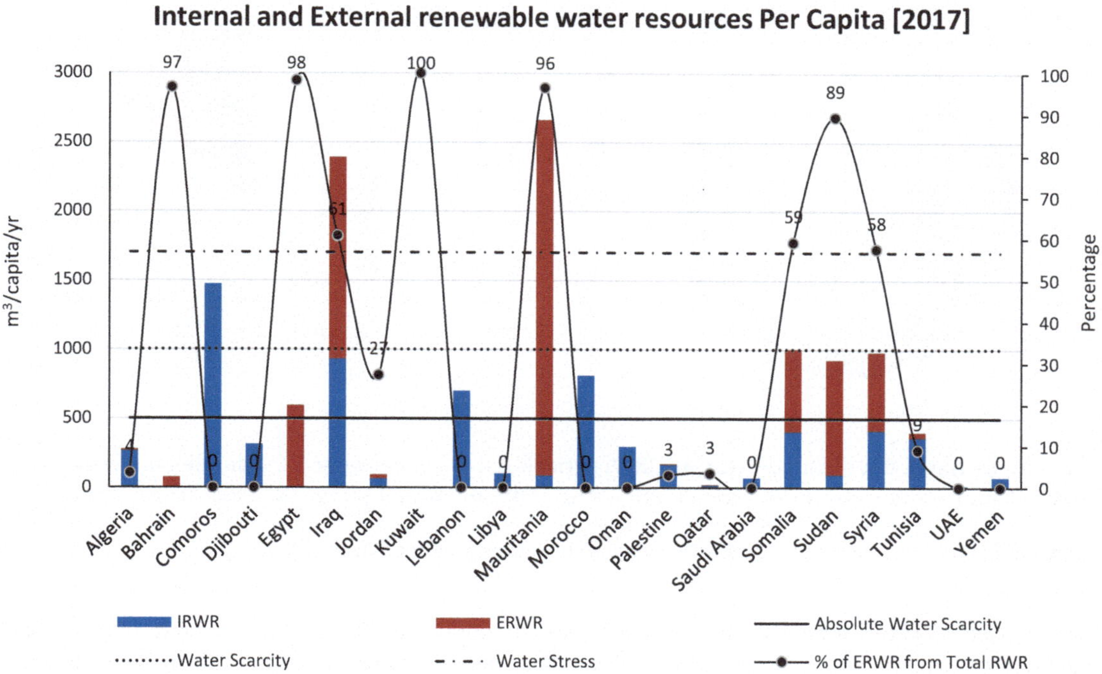

Figure CS13.2 Internal renewable water resources (IRWR) and external renewable water resources (ERWR) in the Arab region (Mohtar *et al.*, 2017).[7]

During the past three decades, water demands in all the Arab countries have increased dramatically as a result of increasing population and urbanization growth, improvements in the standard of living, industrial development and efforts to increase food self-sufficiency. The total water use for all sectors in the Arab region increased dramatically from about 190 billion cubic meters (BCM) in the mid-1990s (ACSAD, 1997) to about 255 BCM in 2010 (UNDP, 2013). The majority of water resources in the region are being used for agriculture (85%), while the municipal and the industrial sectors consume about 8% and 7% of the total water use, respectively (UNDP, 2013; Figure CS13.3).

In the municipal sector, in addition to rapid population growth and urbanization, the rapid increase in urban water demands in the region could be explained by many factors, including rise in per-capita consumption, large losses in the supply network, and lack of recycling programmes within the sector. In many countries, water efficiency in both the supply side and the demand side is generally very low. On the supply side the physical leakage in the municipal networks could reach more than 40%. Moreover, reuse rates of treated wastewater are at their minimal, representing major lost opportunities under the water scarcity conditions of the region. On the demand side, the per capita water consumption in the domestic sector in many countries ranks amongst the highest in the world (e.g., GCC countries).

[7]Internal water resources are those generated within the country, while external water resources are those flowing from outside the country.

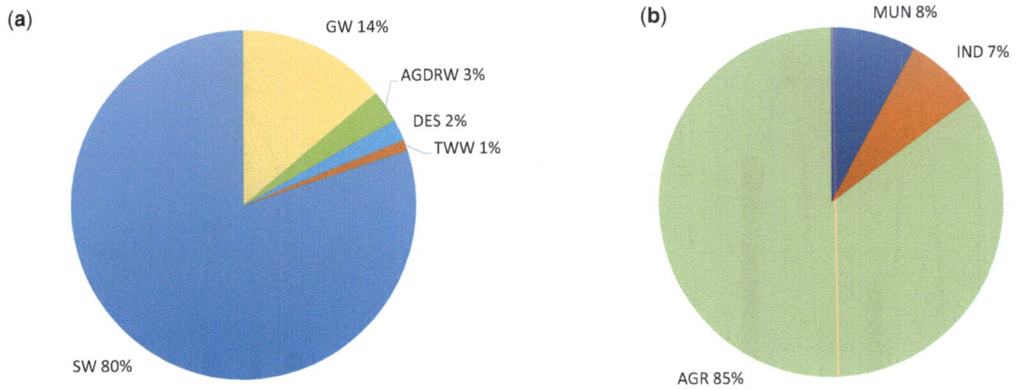

(SW: surface water, GW: groundwater, AGDRW: agricultural drainage water, DES: desalinated water, and TWW: treated wastewater; AGR: agricultural Sector, MUN: municipal Sector, IND: industrial Sector)

Figure CS13.3 Water resources and water uses in the Arab region (UNDP, 2013).

In the face of rising urban demand and the limited supply of conventional water resources, many countries, particularly in the hyper-arid, high-income, energy-exporting countries, such as the Gulf countries, have resorted to desalination. The Arab region leads the world in desalination capacity, where it possesses the majority of the world's desalination capacity; in 2013 total global contracted and online capacity was about 94.5 million cubic meters per day, from which 62.3 million cubic meters per day comes (about 65%) from the GCC countries only (Al Hashemi *et al.*, 2014). Growth in desalination capacity is expected to remain high for the next decade to meet escalating domestic water demand, and the overall share is expected to increase in the region's total water supply as a result of industrialization, accelerated urbanization, population growth, and depletion of conventional water resources. However, desalination is an energy- and capital-intensive process, with water production costs depending on energy requirements, technology growth trends, and environmental impact. However, with recent desalination trends showing improvements in desalination technologies, production costs are dropping. The downward trend in the cost of desalinated water indicates that desalination technology might be becoming more viable for poorer countries. Yet, desalination raises energy security concerns and energy consumption creates a larger carbon footprint, although in GCC countries it takes place mostly in thermal-powered co-generation stations that produce both water and electric power, thereby improving energy efficiency and cost effectiveness (IEA-ESTAP & IRENA, 2012). The GCC countries are also increasingly concerned about the threat to marine life and ecosystems posed by the thermal brines discharged from desalination plants (World Bank, 2012).

CS13.2.3 The food sector

In terms of food, most of the Arab countries have made progress in enhancing their food security situation in the past 10 years, excluding countries in which there was war or civil conflict. Based on the AOAD Arab food security report (AOAD, 2013), domestic production of food commodities increased considerably and is expected to continue to rise. However, some of the Arab countries have over 30% of the population living in conditions classified as poor, 13 Arab countries were classified as low on the Global Hunger Index of 2014, and one country was classified as 'moderate'. Three countries were classified as 'serious', while another three were classified as 'alarming'. Arab countries are unlikely to achieve high ratios of food self-sufficiency; however, they can maintain and improve current ratios (Alzadjali, 2017).

In many Arab countries, agricultural sector performance indicators are very low and agriculture is considered unsustainable due to the continuous deterioration of limited water resources and the limited capacity of arable lands, many having low productivity per unit area. Overall irrigation efficiency is generally low and averages around 45%, while crop productivity is generally low, particularly that of staple cereals, averaging about 1133 kg/ha in five major cereal producers (Algeria, Iraq, Morocco, Sudan, and Syria), compared to a world average of about 3619 kg/ha (Sadik, 2014). Moreover, some countries face serious challenges in their objectives to achieve food security locally. These emanate from a backdrop of constraining factors, including aridity, limited cultivable land, scarce water resources, and serious implications of climate change. Weak policies, insufficient investment in science and technology, and agricultural development have contributed to the impoverished state of agricultural resources and to their inefficient use and low productivity. Population growth, rising demand for food, degradation of natural resources, and conversion of farmland to urban uses pose further challenges to the enhancement of the food security goal in the region.

Moreover, post-harvest losses (PHL) in the region are considered high. It is estimated that the annual losses of grains in Arab countries amounted to about 6.6 million tons in 2012. In addition, loss in imported wheat in some Arab countries translates to about 3.3 million tons due to inefficient import logistics. These national post-harvest losses represent an opportunity cost due to waste of water and energy resources (as well as land and labour resources) used in production (AFED, 2014).

CS13.3 WEF NEXUS DYNAMICS AND RISKS

The complex web of interdependencies between the three sectors and climate variability has manifested itself over the past few years in new and increasingly interconnected crises (the food, energy, and financial crises, together with extreme climate events such as droughts and floods). These crises impacted the Arab population heavily overall and on varying degrees, for example, hitting the poor the hardest.

CS13.3.1 The water–energy nexus

Water and energy are critical resource inputs for economic growth. The risks and the impacts the water sector presents to the energy security and the energy sector presents to water security are numerous. However, in the Arab region these risks are more skewed towards the latter case due to the considerable role the energy sector plays in the water value chain in this arid region, especially for the energy-import dependent countries. Energy inputs are spread across the water supply chain. Energy is used in almost every stage of the water cycle: extracting groundwater, feeding desalination plants with its raw sea/brackish waters and producing freshwater, pumping, conveying, and distributing freshwater, collecting wastewater and treatment and reuse. In other words, without energy, mainly in the form of electricity, water availability, delivery systems, and human welfare will not function. It is estimated that in most of the Arab countries, the water cycle demands at least 15% of national electricity consumption and it is continuously on the rise (Khatib, 2010). Moreover, as easily accessible freshwater resources are depleted, the use of energy-intensive technologies, such as desalination or more powerful groundwater pumps, is expected to expand rapidly leading to more energy consumption.

CS13.3.2 The water–food nexus

The region is already suffering from water scarcity and witnessing intense competition with other sectors, including manufacturing, electricity production, domestic use, and environment. In the face of these competing demands, increasing allocation of water for irrigation will be challenging. Water is a critical input along the different stages of the agri-food supply chain. As the main input in agricultural production, the risks that the water sector presents to food security are considerable and proliferate under the region's arid conditions and sectorial competition. River basins in the Arab region that are critical in the water–food nexus – such as the Nile, Euphrates, and Tigris – are predicted to be 'closed basins' (over-allocated), particularly due to energy and agricultural production, and could face challenges from the effects of climate change and lack of regional water agreements.

On the other hand, agricultural practices have substantial impacts on water security for a broader set of stakeholders. Domestic food security is high on the agenda of many Arab countries. In the wake of the 2008 global food crisis, when at least 25 countries imposed export bans or restrictions on food commodities, many food-importing countries realized the grave food security risks that such situations posed. Several Arab countries (e.g., GCC countries) for which food self-sufficiency is very difficult to achieve, began buying or leasing land in relatively water-rich countries, which by itself creates another risk given that many of the recipient countries are home to significant populations of malnourished and are often on the receiving end of food aid.

CS13.3.3 The energy–food nexus

Agricultural production consumes energy directly in the form of fuels for land preparation and tillage, crop and pasture management, and transportation or electricity supply, and indirectly through the use of energy-intensive inputs, such as fertilizers and pesticides, or energy for manufacturing agricultural machinery. Energy is also needed during processing, distribution, storage, retail, and preparation of food products. This makes food security particularly sensitive to the quality and price of energy inputs: in some countries, the price of oil has a rather direct effect on the price of food (Figure CS13.4). Another dimension of the energy–food nexus that is gaining prominence is the impact of the growing share of

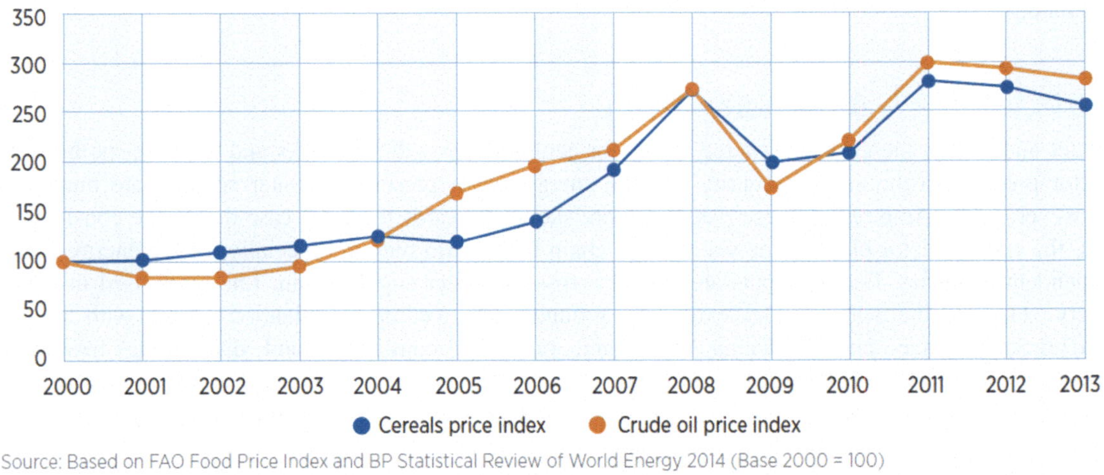

Figure CS13.4 Links and co-risks of world market prices for energy and food: oil–cereal price interlinkages 2003–2013.

modern bio-energy in the world's energy mix, which is emerging as a viable renewable energy option for many countries. For example, liquid biofuels produced from straw or wood and biogas from anaerobic digestion of residues would be the most applicable modern bioenergy options in the region as opposed to traditional sources for example, fuelwood and charcoal.

The nature of energy supply into the agri-food sector can substantially influence food security. The key risk posed by the energy sector on food security is that the dependence on fossil fuels increases volatility of food prices and affects access to food. This risk is magnified in the region's energy-import-dependent countries as fossil fuels continue to provide the majority of the energy inputs for conventional development of the agri-food sector, ranging from electricity and/or diesel for pumping, food processing, and storage, to fuel for agro-machinery.

CS13.4 GOVERNANCE AND INSTITUTIONS IN THE ARAB REGION

It is important to note that, in general, there has been weak or lack of real coordination in the Arab region in terms of integrated policies and strategies for water, agricultural land, and energy. In the majority of the Arab countries, the current water–energy–food–climate policy landscape in the region is fragmented and policies have been developed independently of each other. For example, the current low pricing policies for resources in the majority of Arab countries have been promoting unsustainable consumption and production patterns leading to more resource depletion. Low pricing and across-the-board non-targeted subsidies have resulted in domestic over-consumption of resources and the absence of incentives to achieve resources efficiency.

The institutional framework governing the elements of the WEF Nexus in most of the Arab countries is fragmented, which has in the past and continues today to delay the comprehensive and inclusive management of these interlinked three priorities. This fragmented institutional framework has led to the sectorial approach to policy planning, and consequently fragmented strategies and policies. This fragmentation is also found within the sector itself (e.g., in some countries more than one authority govern the water[8] and energy[9] sectors). Nevertheless, Arab countries present various models of 'integrated institutions;' in the sense one body is responsible for the policy, planning, and/or management of two or more sectors/resources.[10]

It is increasingly evident that development strategies and national policies can no longer be formulated for individual sectors alone. To ensure proper adoption of WEF Nexus, policies and plans must be developed using a multi-stakeholder approach that cuts across the different sectors to address the arising challenges posed by the interdependencies and adequately identify synergies and manage trade-offs. Some Arab countries have a 'higher' water commission providing a level of higher decision making than individual sectors. A good example is the Royal Water Commission in Jordan. The Commission, headed by HRH Prince Faisal Bin Al Hussein, invites representatives of the public and private sector to participate in decision-making processes to ensure a coordinated and holistic approach for water management.

[8]One ministry controls water allocation for domestic and industrial use, while another controls irrigation water use, a third ministry sets standards for potable water quality, and a fourth setting quality standards for surface and groundwater.

[9]For example, energy-related ministries responsible of electricity, renewable energy, and petroleum and minerals.

[10]For example, Ministry of Energy and Water in Lebanon; Ministry of Environment and Water of the UAE responsible for the environment, water resources as well as agriculture, livestock, fisheries, and managing desertification and biodiversity conservation; Morocco combines energy, water, and environment under the Ministry of Energy, Mining, Water and the Environment while the Ministry of Agriculture and Fisheries governs the processes of planning water resources and irrigation.

Uniting key WEF sectors under the umbrella of one ministry is no guarantee for integrated management and governance. Managing the nexus at the local and national level does not require major institutional restructuring, but rather appropriate changes to protocols, procedures, and processes that improve interactions among the relevant ruling entities. The coordination and collaboration mechanisms amongst national institutions are vital elements in applying an integrated approach to resource management. Strong institutions that are better interlinked are means to a nexus approach, and may be more important than additional or new institutions.

For example, many Arab countries have established national cross-sectorial committees for sustainable development[11] and for climate change[12] (Hoff *et al.*, 2017). These committees can be used as key entry points for mainstreaming and implementing the nexus approach by facilitating the integrated implementation of the SDGs and the Nationally Determined Contribution (NDC) of the Paris Agreement.

CS13.5 CONCLUSION

Even though existing institutions of the Arab world face many challenges to a Nexus approach, many opportunities exist that could be tapped into, such as the existing models of integrated institutions and different forms of multi-stakeholders bodies, such as national climate change committees or sustainable development committees could serve as a catalyst to mainstream the Nexus approach at all levels of policy development. Governance and institutional structures in the Arab region can be enhanced and strengthened for more effective and integrated resource management by conducting an in-depth evaluation of institutions and the governance system in each Arab country. This would be for better understanding of the weaknesses and gaps that hinder application of a Nexus approach, and enhancing coordination and collaboration mechanisms amongst the relevant institutions as a key for mainstreaming the WEF Nexus approach, and not necessarily establishing new institutions for the WEF Nexus. The ultimate aim is to have institutions that are able to mainstream and implement the WEF Nexus approach in policies in the Arab countries in light of the mandates and targets of both the SDGs and the Paris 2015 Climate Change agreement and Arab countries' NDCs. This is important in order to ensure that the Arab countries will not, in the near future, be sidetracked by crippling resources insecurities on their sustainable development path.

REFERENCES

ACSAD (Arab Center for the Study of Arid Zones and Dry Lands) (1997). Water resources and their uses in the Arab world. *Proceedings of the First Arab Symposium on Water Resources and their Uses in the Arab World.* Organized by ACSAD, Arab Fund for Economic and Social Development (AFESD), and Kuwait Fund for Arab Economic Development (KFAED), Kuwait City, Kuwait, 8–10 March, 1997, pp. 25–121 (in Arabic).

AFED (Arab Forum for Environment and Development) (2013). Arab Environment 6: Sustainable Energy, Prospects, Challenges, Opportunities. A. Gelil *et al.* (eds.), AFED, Beirut. Available: http://www.afedonline.org/uploads/afed_reports/Sustainable_Energy-English.pdf

---

[11]In Jordan, the 'Higher National Committee for Sustainable Development (HNCSD)' provides a national platform for dialogue on sustainable development issues, challenges and opportunities towards achieving sustainable development goals in Jordan. The Committee includes representatives from 22 different institutions. It is chaired by the Minister of Planning and International Cooperation (MoPIC) and co-chaired by the Minister of the Environment (MoEnv). It has an Executive Secretariat established in the Sustainable Development Division at MoPIC. In Egypt, the 'National Committee to follow up on the implementation of the SDGs' is established under the direct supervision of the Prime Minister and is coordinated by the Ministry of International Cooperation (MIC) in close cooperation with the Ministry of Planning and Administrative Reform (MPAR) and a range of other ministries and councils.

[12]In Egypt, the National Committee for Climate Change is headed by the Ministry of Environment (MOE) and includes representatives from relevant ministries, including ministries of water, energy, and agriculture.

AFED (Arab Forum for Environment and Development) (2014). Arab Environment 7: Food Security, Challenges and Prospects, A.-K. Sadik, M. El-Solh, and N. Saab (eds.), AFED, Lebanon. Available: http://www.afedonline.org/uploads/afed_reports/2014.pdf

Al Hashemi R., Zarreen S., Al Raisi A., Al Marzooqi F. A. and Hasan S. W. (2014). A review of desalination trends in the Gulf Cooperation Council Countries. *International Interdisciplinary Journal of Scientific Research*, **1**, 72–96.

Alzadjali T. M. (2017). Status of food security in the Arab Region. In: Water, Energy, and Food Security Nexus in the Arab Region, Water Security in a New World, K. Amer, Z. Adeel, B. Böer and W. Saleh (eds.), Springer International Publishing AG, Switzerland, pp. 43–57, doi: 10.1007/978-3-319-48408-2_3.

Al-Zubari W. (2017). Status of water in the Arab Region. In: Water, Energy, and Food Security Nexus in the Arab Region, Water Security in a New World, K. Amer, Z. Adeel, B. Böer and W. Saleh (eds.), Springer International Publishing AG, Switzerland, pp. 1–24, doi: 10.1007/978-3-319-48408-2_1.

AOAD (2013). Arab Food Security Report. Arab Organization for Agricultural Development, Sudan.

Doumani F. M. (2008). Climate Change Adaptation in the Water Sector in the Middle East and North Africa Region: a review of main issues. PAP/RAC Workshop, Sardinia, 19–21 May.

ESCWA and UNEP (2015). Arab Sustainable Development Report 2015. ESCWA report number E/ESCWA/SDPD/2015/3.

ESCWA *et al.* (2017). Arab Climate Change Assessment Report – Main Report. United Nations Economic and Social Commission for Western Asia (ESCWA), RICCAR (Regional Initiative for the Assessment of Climate Change Impacts on Water Resources and Socio-economic Vulnerability in the Arab Region). Beirut, E/ESCWA/SDPD/2017/RICCAR/Report.

Falkenmark M. (1989). The massive water scarcity now threatening Africa – Why isn't it being addressed. *Ambio*, **18**, 112–118.

Fattouh B. and El Katiri L. (2012). Energy and subsidies in the Arab region. UNDP Arab Human Development Report Research Paper Series, Regional Bureau for Arab States, New York, NY.

Hoff H., Mansour L., Carius A., Kramer A., Abaza H., Al Ouran N., Al-Zubari W. and Ulrich A. (2017). Mainstreaming the Water-Energy-Food Security Nexus into Policies and Institutions in the MENA Region: National guidelines on mainstreaming the water-energy-food (WEF) security nexus into policies and institutions in Egypt and Jordan. GIZ, GFA.

IEA-ESTAP (International Energy Agency – Energy Technology Systems Analysis Program) and IRENA (International Renewable Energy Agency) (2012). Water desalination using renewable energy, Technology Brief, No. 112 (March 2012). Available from: www.irena.org/DocumentDownloads/Publications/IRENAETSAP%20Tech%20Brief%20I12%20Water-Desalination.pdf.

Khatib H. (2010). The Water and Energy Nexus in the Arab Region. Unpublished consultancy report, League of Arab States, Cairo.

Khody, K. and Gitonga, S. (2017). Status of Energy in the Arab Region. In: Water, Energy, and Food Security Nexus in the Arab Region, Water Security in a New World, K. Amer, Z. Adeel, B. Böer and W. Saleh (eds.), Springer International Publishing, Switzerland, pp. 25–41, doi: 10.1007/978-3-319-48408-2.

Koday K. (2012). Sustainable Development as Freedom: Energy, Environment and the Arab Transformation. Poverty in Focus for Inclusive Growth, Vol. 23. International Policy Center, Brasilia.

Mohtar R., Assi A. and Daher B. (2017). Current water for food situational analysis in the Arab Region and expected changes due to dynamic externalities. In: Water, Energy, and Food Security Nexus in the Arab Region, Water Security in a New World, K. Amer, Z. Adeel, B. Böer and W. Saleh (eds.), Springer International Publishing AG, Switzerland, pp. 193–208, doi: 10.1007/978-3-319-48408-2_10.

RECREEE (Regional Center for Renewable Energy and Energy Efficiency) (2015). Arab Future Energy Index: AFEX 2015, Energy Efficiency. RECREEE, Cairo.

Sadik A. (2014). The state of food security and agricultural resources. In: 2014 Report of the Arab Forum for Environment and Development, Arab Environment 7: Food Security, Challenges and Prospects, A. Sadik, M. El-Solh and N. Saab (eds.), Arab Forum for Environment and Development (AFED), Beirut, pp. 12–43.

Sdralevic C., Sab R., Zouhar Y. and Albertin G. (2014). Subsidy reform in the Middle East and North Africa. IMF, Washington, DC.

UNDESA (United Nations Department of Economic and Social Affairs) (2015). World Population Prospects: The 2015 Revision; [http://esa.un.org/unpd/wpp].

UNDP (2013). Water Governance in the Arab Region: Managing Scarcity and Securing the Future. UNDP Regional Bureau of Arab States (RBAS). [Available at: http://www.arabstates.undp.org/content/rbas/en/home/library/huma_development/water-governance-inthe-arab-region.html].

World Bank (2012). Renewable Energy Desalination: An Emerging Solution to Close the Water Gap in the Middle East and North Africa. World Bank, Washington, D.C [Available from: http://water.worldbank.org/sites/water.worldbank.org/files/publication/water-wpp-Sun-Powered-Desal-Gateway-Meeting-MENAs-Water-Needs_2.pdf].

Water–energy–food Nexus and environment in Central Asia

S. Xenarios[1], V. Smakhtin[2], J. Sehring[3], D. Schmidt-Vogt[4], S. Tsani[5,6,7], C. Hannah[8] and E. Michalena[9,10,11]

[1]Nazarbayev University, Graduate School of Public Policy, Astana, Kazakhstan
[2]United Nations University, Institute for Water, Environment and Health (UNU-INWEH)
[3]IHE Delft Institute for Water Education, Integrated Water Systems and Governance Department, Delft, The Netherlands
[4]Freiburg University, Faculty of Environment and Natural Resources, Chair of Silviculture, Freiburg, Germany
[5]Athens University of Economics and Business, Athens, Greece
[6]International Centre for Research on the Environment and the Economy, Marousi, Greece
[7]Centre for Euro-Asian Studies, University of Reading, Henley Business School, School of Economics, Reading, UK
[8]Nicholas School of the Environment, Duke University, Durham, USA
[9]University of the Sunshine Coast, Australia
[10]Centre of Renewable Energy (CORE), Fiji
[11]Sorbonne University and CNRS, France

ABSTRACT

Transboundary river basin systems comprise the backbone of the WEF Nexus in Central Asia (CA). In Soviet times, the water-abundant upstream countries provided water for irrigation in summer to downstream countries, and, in exchange, were supplied with energy in winter. Environmental aspects of natural resources management were neglected. The desiccation of the Aral Sea is the most widely known example of the consequences of disregarding environmental impacts. After independence in 1991, the formerly regional approach to water and energy management was replaced by a national approach which led to fragmentation of supply networks and conflicts mainly between upstream and downstream neighbours. This study outlines interventions and technologies for the water, energy, agricultural, and environmental sectors at various scales that would be suitable for the introduction of a sustainable regional WEF in CA. In the water sector, designing hydropower systems in upstream countries should consider the release of sufficient water volume for downstream irrigation in summer while the pressure on surface river basins should be mitigated through the increased use of groundwater resources. The energy portfolio in all countries of CA should be diversified through more investment in renewables, and by designing the macroeconomic perspectives of energy policies in the region so as to avoid boom and bust pricing fluctuations on energy commodities. In the agricultural sector, the horticultural and tree cultivation should be prioritized over cotton monoculture. In addition, the rehabilitation of irrigation and drainage systems should be conducted at operational, maintenance, and investment costs that are affordable to farming communities. Providing minimum environmental flows in major river networks and the better monitoring of human activities on basin level are prerequisites for an integrated approach of the WEF Nexus in CA.

Keywords: WEFE Nexus, hydro-climate, hydropower, agriculture, Aral Sea

CS14.1 INTRODUCTION

Central Asia (CA) is a region where water, energy, food, and environment are inextricably linked. This is due to: (i) diverse ecosystems ranging from glaciers, high mountain areas, forests, to oasis regions, and vast steppes and deserts associated with spatial water resources variability; (ii) the structure of the economy, traditionally dominated by agriculture and supported by large-scale hydropower and irrigation schemes, and (iii) relatively recent geopolitical changes which led to fragmentation of a formerly contiguous region into new nation states separated by international boundaries, creating a new reality and new challenges for managing natural resources, water, and energy – in particular.

The upstream countries of Afghanistan, Tajikistan, and Kyrgyzstan are located in the mountains of Karakoram, Pamir, and Tien Shan, often referred to as 'water towers' of CA. Snowfall and glacier melting are important sources of water in an overall dry climate, feeding into a vast river network, which traverses the entire region. The downstream countries of Turkmenistan, Uzbekistan, and Kazakhstan are situated on extensive plains, largely converted from grasslands to agricultural lands, which are also endowed with abundant hydrocarbon (coal, oil, gas) resources.

All CA countries, except Afghanistan, were part of the Union of Soviet Socialist Republics (USSR), and have gone through a period of large-scale engineering solutions to water resources development, with a focus on surface water management for irrigation and hydropower. The USSR pursued a policy of regional development with specialization in each region on one product or a small range of products which were distributed through internal exchange mechanisms. This led to monoculture development, for example, the development of huge cotton plantations in downstream areas, with little concern for the environmental impacts. Upstream countries, where most of the CA river flow is generated, were providing water for irrigation to downstream countries. In exchange, they were supplied with coal and gas for heating and electricity purposes in cold months as well as with agricultural products and especially staple crops.

Intensification of agricultural production, regulation of waterways through dams and other infrastructure and over-abstraction of surface water in the entire CA came at a high price to the environment, resulting in the well-known impacts on the Aral Sea. The fragmentation of the region into several independent states in the late 1990s changed the approach to the WEF Nexus from regional exchange to country-focused policies with each country developing at its own pace and pursuing different development objectives. As a consequence, there were changes in the water use for energy, food, and environment which have led to tensions mainly between the upstream and downstream states.

The transboundary nature of water resources in CA remains of paramount significance for the development of the energy, water, and agricultural sectors of the region as a whole, creating interdependencies that should, eventually, inspire and strengthen regional cooperation. The objective of this paper is to examine how a sustainable regional WEF Nexus approach can emerge in CA in the current and developing geopolitical circumstances, and which appropriate WEF interventions and technologies may best support the development needs of each CA country. Each WEF sector is assessed separately on a regional scale while the natural environment is also acknowledged as a separate component.

CS14.2 STATE OF THE ART

CS14.2.1 Transboundary water systems

The two main rivers of the region, the Amu Darya and Syr Darya, are the most important water sources for the livelihoods of about 70 million inhabitants in CA. Both rivers originate in the mountain ranges of the upstream countries. The Amu Darya, the largest river in CA in terms of water volume, is formed by the Panj River on the Tajik-Afghan border and the Vakhsh River in Tajikistan and continues into Uzbekistan and Turkmenistan before emptying into the Aral Sea. The Syr Darya, which is the longest river in CA, has its source in the Tien Shan Mountains in Kyrgyzstan and flows through the Fergana Valley into Tajikistan and Uzbekistan and then into Kazakhstan where it ultimately discharges into the Aral Sea.

Central Asian countries are among the most water-intensive economies in the world with a mean water withdrawal per capita at 2200 m/year and more than 80% of water used by the downstream countries, mainly for irrigation (Sehring & Diebold, 2012). Irrigated farming in CA was strongly prioritized in Soviet times through the construction of numerous reservoirs, extended supply and drainage networks and large pumping stations. Water management was organized according to 'water-use regions' or 'irrigation districts', which in some instances were transgressing the republican boundaries (Wegerich *et al.*, 2012). The post-Soviet breakdown has differently impacted each district depending on the facilities, organizational structure, and other parameters by avoiding a collapse of the entire agricultural water supply network.

The transboundary irrigation districts ceased to exist after independence in 1991 and administrative boundaries were drawn instead for water resources management on the national level. Intensified irrigation practices continue until today although at a slower pace due to the lack of funding for maintenance and operational services (Dukhovny & de Schuetter, 2011).

In the last decade, a river basin management approach was gradually introduced in each of the CA countries in an attempt to improve national water use and allocation plans as per the principles of the European Water Framework Directive (WFD, 2000/60/EC). In particular, the European Union Water Initiative was established in 2002 as transnational, multi-actor partnership to support water governance reforms on the globe (Fritsch *et al.*, 2017). For the region of Eastern Europe, Caucasus and CA, one partnership was established with 12 countries for the improvement of the legal and regulatory water-related frameworks in alignment with WFD, development of River Basins Management Plans (RBMPs) and engagement of stakeholders through National Policy Dialogues (NPDs) and River Basin Councils (RBCs).

For the implementation of the RBMPs, River Basin Organizations (RBOs) have to be established to monitor all activities related to water management on a basin level. Indicatively, the RBMPs have to gather information from local (e.g., WUAs) and centralized institutions (e.g., Ministries), reduce the currently unregulated water withdrawals from rivers, canals, and newly built groundwater wells which are still common practice in rural areas. All CA countries have set up the legal basis for introducing the basin approach while RBMPs have been developed gradually in Kazakhstan, Kyrgyzstan, and Uzbekistan over the last 3 years (2014–2017). The legislative documents, mostly acknowledged as Water Codes, emphasize the WEF Nexus of each country and the need to harmonize the respective national documentation with that of the neighbouring countries. In practice, the Water Codes dictate how various uses (drinking, agricultural, industrial, environmental) will be regulated and prioritized according to different needs and set the foundations for a comprehensive management of water resources within the country. They also set provisions for water flow requirements of major river systems in respect to international treaties and conventions.

The RBMPs in the upstream countries will be obliged to monitor the construction of big controversial dams such as the Roghun in eastern Tajikistan and Karambata I in north Kyrgyzstan for hydropower production. These large-scale interventions have up until recently created conflicts between the upstream countries and Uzbekistan as the most vocal downstream country. The current and future hydropower developments are presented in Figure CS14.1.

Tensions in the region have abated somewhat, as a result of feasibility studies (SNC-Lavalin, 2017) that were conducted for both dams, and of Strategic Environmental Assessments (World Bank, 2014) (SEIAs) to evaluate the impact on downstream water volume, but more importantly because of the recent change (2016) in the government of Uzbekistan.

Moreover, investments in technological interventions in oil, gas, coal, and uranium mining in the region are slowly increasing in volume (Kazakhstan Green Energy, 2017). The interventions are anticipated to reduce water input and also mitigate the water pollution induced by these activities. Further, international organizations and donors are investing in the rehabilitation of irrigation networks and drinking water supply mainly in rural areas of Tajikistan, Kyrgyzstan, and Uzbekistan.

However, there is a notable lack of coordination, monitoring, and assessment of these interventions which is mainly due to the overlapping between too many governmental authorities and the differences in prioritizing water resources management by the governments. For instance, Kazakhstan sets a high priority on water management for food production and holds the Ministry of Agriculture responsible for the development and implementation of agricultural policy and water management. Groundwater use remains under the supervision of the Ministry for Investment and Development and the Committee of Geology and Subsoil Use (UNECE, 2017). A similar situation presents itself in Uzbekistan where the Ministry of Agriculture and Water Resources is responsible for surface water resources and the State Committee on Geology and Mineral Resources for ground water. Agriculture also plays a dominant role in downstream Turkmenistan where the Ministry of Agriculture and Water Resources is mainly accountable for efficient agricultural water management.

Kyrgyzstan has attempted to assign greater importance to the water sector by establishing the National Water Council (NWC) in 2005 which has the responsibility to coordinate all the state and private agencies involved with water resources management. In reality, the NWC has remained inactive for many years and the newly introduced basin approach is in substance implemented by the Department of Water Economy and Melioration under the Ministry of Agriculture and Melioration. The clear priority on water for energy use in Tajikistan found expression in the creation of the Ministry of Energy and Water Resources (MEWR) in 2013. While hydropower development is the primary mandate of the Ministry, water for agricultural use is supervised by the Agency of Land Reclamation and Irrigation (ALRI), which is of inferior importance to the Ministry. Coordination of the activities of these authorities by the national RBOs and communication of these RBOs within the framework of a river basin (e.g., Syr Darya) for a more balanced WEF Nexus remain a major challenge to be confronted in the CA region.

CS14.2.2 Energy sector and trends in national policy frameworks

The CA region looks back on a long history of energy resource abundance (oil, coal, and gas in Kazakhstan, gas in Turkmenistan and Uzbekistan, hydroelectric power potential in Kyrgyzstan and Tajikistan). Following the break-up of the Soviet Union in the 1990s, CA countries transited to a decentralized

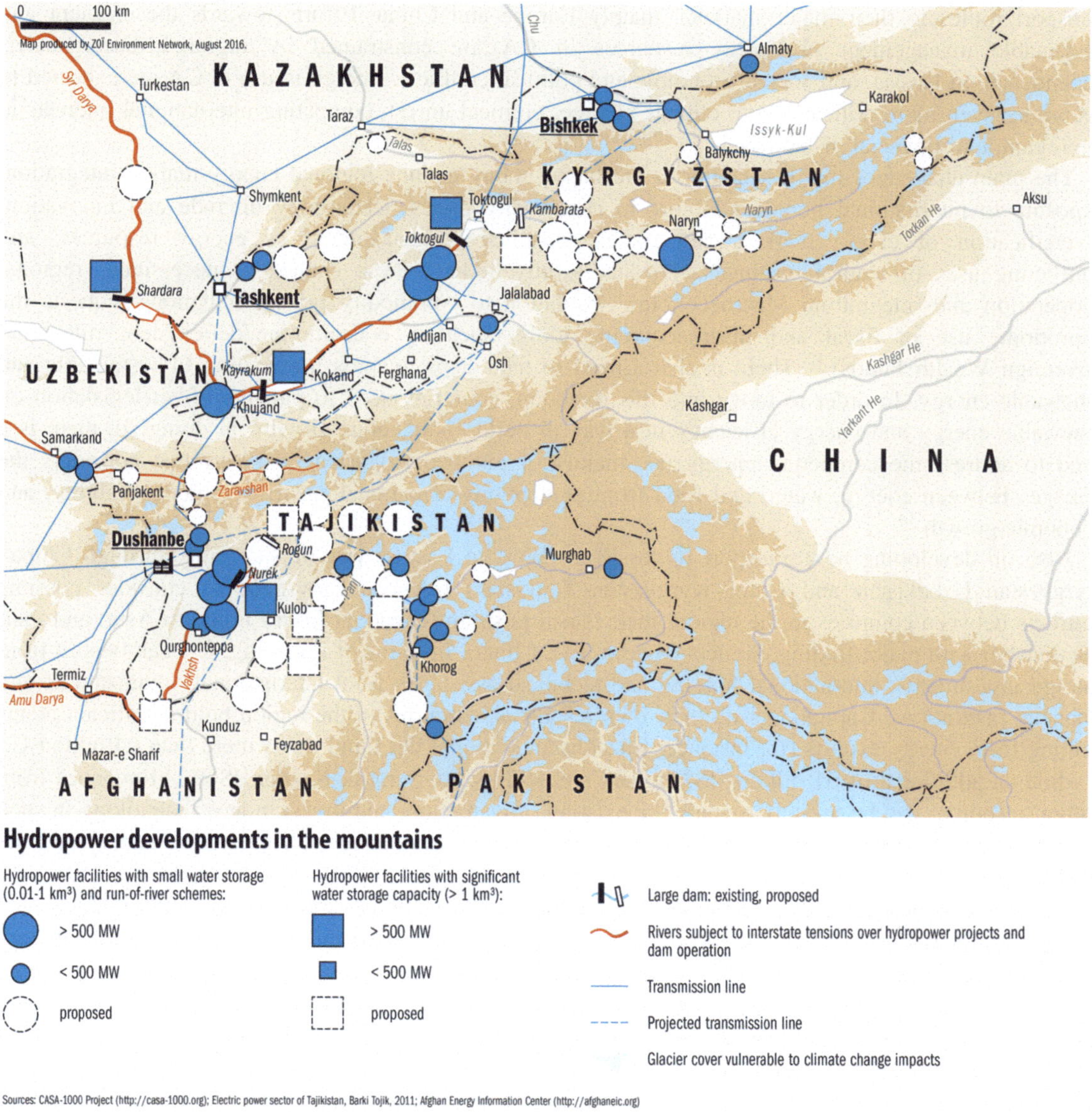

Figure CS14.1 Hydropower developments in CA, with focus on the upstream countries of Kyrgyzstan and Tajikistan [in page 65, Zoi Environmental Network], 2016. The names and boundaries shown and the designations used on this map do not imply official endorsement or acceptance by the European Commission or the United Nations. *Source*: [https://www.osce.org/ secretariat/355471].

energy production and management system that has been driven by national goals and capacities regarding technological state and advancement, exports, explorations, and production (EBRD, 2004). The transition period has been characterized by decaying and dilapidated infrastructure and technology, decoupling transmission and distribution systems and poor energy policies at macroeconomic level. In the last decades, CA countries have aimed at exploiting in full their energy potentials and at developing new transport routes to their major markets, mainly Europe and China. Efforts towards the efficient and sustainable management of energy resources in CA are constrained by regional discrepancies, dependence on Russia, and social, economic and political conflicts. Energy issues in CA are expected to becoming even more critical due to climate change, political unrest, competing international interests in the region, and financial instability.

The main challenges that CA countries are facing today include regional energy market integration, exposure to international boom and busts in commodity prices, technology upgrade and innovation, diversification of export markets and routes, sustainable management of energy resources, and conflicting uses of natural resources. To successfully address these challenges necessitates regional cooperation and integration. Macroeconomic insulation to commodity price fluctuations requests the appropriate use of fiscal and macroeconomic tools (such as consumption/deficit/debt rules and Sovereign Wealth Funds).[13] There is also a need to take steps towards a transition to sustainable and renewable energy. In order to move into this direction, countries need to put into effect legislation on renewable energy and energy efficiency in a way to build endogenous growth. In doing so, there is a need to address the competing energy use mainly of pumped irrigation in a way that considers the linkages between energy, water, and food from perceptions of national security, regional stability, and economic growth.[14]

Also, in developing new transport routes (from West to East or vice versa mainly through China–Kyrgyzstan–Uzbekistan and China–Kyrgyzstan–Tajikistan) CA countries need to consider possible conflicts between countries in the region, their transit position (corridors from Europe to Asia) and their relationship with major trading partners (Russia, EU, China). In terms of financing, CA can benefit from foreign investments, international funding and technical assistance from international organizations and financial institutions. Capitalizing on their own resources, countries in the region hold significant assets coming from energy exports (for instance, Kazakhstan National Oil Fund holds more than 64 billion US $ while an additional of 61 billion US$ dollars are held by the Samruk-Kazyna JSC).[15] Revenues from natural resources can be used to upgrade the existing infrastructure and invest in new technology in such way so as to address the long-term challenges CA countries are facing with regards to energy and the WEF Nexus.

In terms of regional energy collaboration, water and energy resources were managed during the Soviet period – as noted above – in an integrated and top-down approach. After independence, the unified energy system broke down, and downstream states demanded market prices for their energy fossils. In 1998, Kazakhstan, the Kyrgyz Republic, and Uzbekistan signed an 'Agreement on the Use of Water and Energy Resources of the Syr Darya Basin', later joined by Tajikistan as well. This agreement provided that Kyrgyzstan would discharge water from its reservoirs in summer for the downstream states of Kazakhstan and Uzbekistan, while these would deliver fuel to Kyrgyzstan in winter, so that the country would not need to rely on hydropower. The agreement required annual protocols to define exact discharge times and amounts, as well as the price of energy to be sold to the downstream countries during the summer period and on the transfer prices of coal, gas, and electricity. The agreement worked well for some years, but from 2003 on, the parties failed to agree on annual protocols. Efforts were made to deal with this problem, but a lasting solution has not yet been found.

---

[13]See: Tsani (2013), Tsani *et al.* (2011), Ahmadov *et al.* (2011), Kalyuzhnova (2011).

[14]See among others Keskinen *et al.* (2016), Karatayev *et al.* (2016), Freedman and Neuzil (2016).

[15]See: http://www.swfinstitute.org/sovereign-wealth-fund-rankings/

CS14.2.3 Food security and agriculture

Food and agriculture systems significantly contribute to the Central Asian region's economy and food security. The agricultural sector makes up roughly 15–20% of the national GDP for Afghanistan, Kyrgyzstan, Tajikistan, and Uzbekistan while Kazakhstan and Turkmenistan depend less on the agricultural sector (5% and 12%, respectively) due to opportunities in the oil and natural gas industries (Pomfret, 2006). Agriculture provides employment to about 23 million people in CA, most of whom are residing in rural areas. The upstream countries are more engaged in farming activities. For instance, in the least developed countries of Afghanistan and Tajikistan about 79% and 53%, respectively, of the entire labour force are mainly occupied in subsistence farming while only 18% of the labour force in Kazakhstan contributes to the agriculture sector, which is dominated by commercial farming. The agricultural sector of CA is mainly consisted of irrigation farming, followed by pastoral and aquaculture/fisheries production. All these activities contribute to local livelihoods and depend on water and energy resources to sustain food production.

Over the 20th century, intensive cotton production dominated the Central Asian economies with pumped (or lift) irrigation as the main driver of intensified cultivation. Today, pumped irrigation occupies nearly half (44%) of the arable land in Tajikistan, about one third (27.4%) in Uzbekistan, and a smaller but still noteworthy (16.3%) area in Turkmenistan. Cereals, vegetables, and orchard trees have surpassed cotton production since early 2000, especially in Kazakhstan but lately also in Uzbekistan and Tajikistan (Frenken & Food and Agriculture Organization of the United Nations, 2013). However, the transition from intensive and water-demanding monocultures to more diversified production has not provided the anticipated relief on irrigation pressures and freshwater is still being extracted at alarming rates.

In Turkmenistan cultivations suffer from water-stress conditions the most[16] (107%) followed closely by Uzbekistan (100%) then Tajikistan (65%), and Afghanistan (31%). Water appropriation to croplands is also seriously affected by significant irrigation water losses all over CA. For example, only an estimated 30–35% of the initially lifted water is delivered to croplands in Tajikistan while the rest is wasted due to inefficient water-use practices and decaying irrigation infrastructures (Abdullaev *et al.*, 2010). The pastoral systems of CA's steppe and mountain landscapes, as well as fisheries/aquaculture in irrigated areas and downstream countries, are also inextricably linked to water resource systems. The availability of water and the type of water sources (e.g., spring, wells) have determined the movement of livestock across the region (Rahimon, 2012). Irrigated water contributes to fodder production for feeding livestock, as well as aquaculture production. Freshwater resources are also required to support downstream fishery systems, which are negatively affected by water pollution and shortage due to irrigation, up to the total disappearance of fish in the Aral Sea. On the other hand, in the Northern Aral Sea fishery could be revived thanks to rehabilitation of fish stocks after the construction of the Kok-Aral dyke and to water-saving measures along the Syr Darya River, which led to the increase in water volume and quality in this separated part of the Aral Sea (Pala, 2011).

Major challenges to sustain agriculture systems in CA include monitoring and controlling the tremendous water losses in the irrigation sector, financing and facilitating irrigation infrastructure rehabilitation efforts to support agricultural production, promoting energy-efficient solutions for pumped irrigation systems, and providing information and trainings for people engaged in the agriculture sector on water resource use and local-level climate change adaptation strategies that are relevant to food production.

[16]A country is water-stressed if populations withdraw more than 25% of their renewable freshwater resources, (2) approaching physical water scarcity when more than 60% is withdrawn, or (3) facing physical water scarcity when more than 75% is withdrawn (IWMI, 2008). Uzbekistan and Turkmenistan are two of the 10 countries in the world that withdraw more than 100% of their renewable water resources for irrigation, which means they also withdraw water from groundwater sources.

Imminent action should be taken on technical, institutional, and policy aspects of making farming economically viable in CA. For example, the decoupling of high-energy intensity pumping and irrigation could be achieved by the installation of more efficient and renewable energy pumping systems, such as solar pumping (IRENA, 2016). Local water management institutions like WUAs would greatly benefit from greater authority and recognition from state government agencies, as well as targeted funding to better monitor and assess local level irrigation infrastructures conditions and water-use practices. At the policy level, agricultural development objectives in CA should also take into consideration the linkages between irrigation infrastructure rehabilitation efforts and its economic potential for improving the welfare and livelihoods of CA's populations.

CS14.2.4 Status of the environment in CA

The management of natural resources in CA has continued to pursue the 'nexus' approach of Soviet times through integrated management of water, land, and energy resources. Based on the belief engineering and technical supremacy over nature, environmental aspects were neglected which entailed vast consequences especially for freshwater resources (Granit *et al.*, 2012). The desiccation of the Aral Sea, 80% of which have turned into desert, has emerged as a symbol of unsustainable water management in the Soviet era. Much of the irrigated land is plagued by salinization, waterlogging, and water erosion. The inflow of drainage water is heavily contaminated with nitrates, organic fertilizers, and phenol which eventually also pollute surface and groundwater sources. Uranium legacy sites, most of them un-remediated, poorly secured, and often located near Transboundary Rivers and in disaster-prone locations, constitute a continuing threat. In addition, the degradation of wetlands, desertification, land degradation, erosion, and increasing landslides, are some of the major environmental challenges to be met nowadays in the CA region. Climate change is bound to increase pressures on the environment and to augment the vulnerability of ecosystems and livelihoods (Alfrord *et al.*, 2015).

The linkage of the WEF Nexus with the natural environment in CA is heavily pronounced in all three sectors of water, energy, and food. The over-abstraction of surface and groundwater has significantly decreased the flow in all major rivers with severe impacts on the fauna and flora mainly in downstream areas. Efforts to replenish aquifers in the Syr Darya and Amu Darya basins have so far met with limited success (Karimov *et al.*, 2010). The outdated irrigation network and the dilapidated drainage systems in many agricultural areas causes soil degradation and salinization problems. Maintenance of such infrastructure is especially precarious in transboundary areas where border demarcation is still in process and willingness to invest is low. Water quality has also deteriorated from insufficient or lacking waste water treatment in domestic and industrial use (Abdullaev & Rakhmatullaev 2016; UNEP, UNDP, UNECE, OSCE, REC & NATO, 2011).

The energy sector also contributes strongly to the degradation of the natural environment in CA. The most noteworthy impact is from pollution by the mining industry in the entire CA region. Gold mining for instance, allegedly contaminates surface waters in eastern Kyrgyzstan while coal mining has caused soil and water contamination by chemicals used in mining processes in all three downstream countries (ZOI, 2013). The regulation of river flows in upstream countries through the construction of cascading reservoirs and large dams has induced erosion and soil degradation which also affects water quality.

The current economic development trends in CA can have mixed impacts on the environment. In agricultural and drinking water supply, recent and ongoing technical investments and institutional reforms have led to some improvement in water-use efficiency which can relieve water stress in the entire region. In the energy sector, Kazakhstan, Turkmenistan, and Uzbekistan, despite investment programmes for alternative energy sources, still rely heavily on fossil fuels. The development of new hydropower stations remains a priority for Kyrgyzstan and Tajikistan, which produces clean energy but nevertheless has environmental impacts on the river ecosystems.

Of late, the concept of 'Green Economy' has gained prominence in CA, most specifically in Kazakhstan, which in 2014 adopted the 'Concept for the Transition to a Green Economy'. Among others, the concept includes provisions for more sustainable water usage, increase in protected areas and fostering of low-carbon energy. Within the Green Economy framework, both Kazakhstan and Kyrgyzstan have stated the provision of a minimum environmental flow in Syr Darya and the Northern Aral Sea (UNECE, 2017). Although the concrete effects of the Green Economy initiatives are still to be seen, they can provide a framework for implementing the nexus approach and link it to issues for which there is political will.

While intersectorial governance and coordination remain a challenge both at the national and regional levels, there are some existing arrangements that allow addressing environmental aspects in a nexus approach in CA. The *Interstate Commission for Sustainable Development* (ICSD) was established in 1994 as a sub-body of the International Fund for Saving the Aral Sea (IFAS), which is the only formal organization facilitating cooperation on environmental protection and sustainable development CA. Given this structure, the ICSD is focussed on water–environment linkages and in particular the impacts of agriculture on these. It could, if strengthened, be a good platform to co-ordinate efforts to tackle water–agriculture–environment linkages, while the links to energy, as in the whole IFAS structure, are missing. At the national level, all CA countries possess relevant environmental strategies and legal frameworks, and have ratified relevant international agreements. Their implementation and compliance however still faces many challenges.

CS14.3 KEY FINDINGS

Following the structure of this paper, the key findings for each sector as well as for the environmental component will be presented in a stepwise manner. The key finding suggests that the transboundary nature of water resources in CA constitutes the major element of the WEF Nexus and most likely also of economic development in the entire region. This is due to geophysical reasons but also partly attributed to the high water, energy, and food interdependence of CA during the Soviet period which is still seen up to today. Significant attempts have been made to reduce energy losses in agriculture through the rehabilitation of irrigation systems, changing of monoculture to more economically valuable and less water-demanding crops while the utilization of hydropower potential upstream is meant to be conducted in line with agricultural needs in the downstream counties. On the other hand, lift irrigation schemes which are not so reliable and which are heavily dependent on energy from the grid and old and derelict USSR infrastructure are still in operation in all CA countries. In the downstream countries of Uzbekistan and Tajikistan new large pumping stations are still constructed according to the Soviet model. There is as yet very little usage of groundwater irrigation and also limited research has been on the conjunctive management of surface and groundwater sources (Karimov *et al.*, 2015). Drinking water supply is also still heavily dependent on surface water systems. Studies have shown that the countries of CA, especially the downstream countries, have the potential to use their groundwater reserves more efficiently and relieve the pressure on riverine systems (Karimov *et al.*, 2015). This relief from pressure could also lower the tension over water sources for energy between upstream and downstream countries and encourage a more even allocation of water resources in the region.

In case of the energy sector, the region has to deal with ageing infrastructure, a low technological readiness level, high-energy intensity, low-energy efficiency, untapped renewable energy potential, poorly functioning regional energy markets, conflicting use of natural resources (consider the case of water for energy and/or agriculture), environmental pollution and ecosystem degradation and a poor legal and institutional framework for regional cooperation, particularly with regards to the exploitation and management of shared resources. Many of these problems are expected to augment as a result of climate change. In the context of regional sustainable development, the transition towards renewable energy should be put at the top of regional cooperation and national energy policy priorities. For a successful and efficient transition towards renewable energy and green growth, a special attention to the WEF Nexus should also be paid. The commodity price fluctuations in the last two decades, that coincided with the outbreak of the last economic crisis, and the recent development of alternative energy resources (shale gas in USA) made clear to energy exporters in the region the challenges of managing volatile natural resource revenues. Energy exporters in the region need thus to consider carefully the use of fiscal and macroeconomic tools so as to address similar problems in the future.

The agricultural and food sector in CA is especially dependent on freshwater resources to sustain production. Although commendable water-saving efforts have been made in recent years to move away from water-intensive cotton monoculture crop production to less water-intensive crops, countries in CA remain some of the most water-stressed countries in the world. In addition, melting glaciers and increasing aridity may also affect the future availability of freshwater resources. Under these conditions, it is especially important that WEF policies in the CA region help to develop better monitoring and assessment of water losses, support the rehabilitation of gravity-fed and pumped irrigation infrastructures, identify energy-efficient opportunities for pumped irrigation systems, reconsider their water allocation to agriculture and provide guidance on climate change adaptation strategies that are most relevant to local communities. Local level institutions, such as WUAs, which provide a link between local communities and state government agencies, can play a major role in facilitating these types of agricultural-based WEF policies. However, for these WEF policies to be most effective, it is important for state government agencies in CA to recognize the impactful role that local institutions can have in monitoring irrigation infrastructures conditions and water-use practices.

Finally, in terms of the natural environment, CA undeniably faces catastrophic environmental consequences due to the neglect of environmental aspects in earlier (and current) water, food, and energy policies. These negative environmental impacts are not only a result of trade-offs between sectors (which also negatively affect interstate relations), but also partly due to the earlier integrated approach (see above). These devastating impacts are not only an environmental concern, but also threatened the quality and availability of resources like freshwater and land, which are crucial for economic development across sectors, be it in agriculture, energy, or industry.

It is therefore essential that any effort to promote the WEF Nexus approach in CA takes the environment into account. A water–energy–food–ecosystems/environment Nexus approach is needed for development and stability in the region. Implementing a nexus approach to the transboundary management of water, land, and energy resources can yield concrete benefits for CA, such as improved water quality and more reliable access to water, improved status of ecosystems and better ecosystem services, reduced GHG emissions and increased resilience towards disasters and climate change. Such an approach benefits the environment as well as the economies and societies, and would fit into green economy strategies of the countries.

CS14.4 CONCLUSIONS AND RECOMMENDATIONS

The development of national agricultural and energy resources in CA is strongly dependent on transboundary water sources. The interdependence of CA countries with respect to water underlines the need of a regional WEF Nexus framework. However, each country's focus is on national policies and objectives rather than regional perspectives. This is further accentuated by the fact that the various countries occupy different development stages. The development disparities between the individual CA countries are likely to encourage frictions and conflicts over water management which is the pivot of economic development both in upstream and downstream countries.

An enabling precondition for strengthening the Nexus approach in CA is that, given the legacy of the water–food–energy policies of the Soviet Union, integrated approaches were in place to tackle these issues in the region even before the 'Nexus' term gained prominence. Current examples of different initiatives that tackle water, food, and energy issues in an integrated way are the SPECA[17] programme as well as activities by international funding agencies, development banks, and multilateral organizations. Such policies are therefore not entirely new, but can be built on earlier approaches.

At the political level, the Nexus concept can provide an opportunity to overcome the deadlock the region faces in transboundary and intersectorial water management. The water discourse is highly politicized in CA. Approaches which directly aim at changing existing water-use patterns risk to be faced with mistrust and opposition by ending in the 'water trap' (Abdullaev & Rakhmatullaev, 2016) notion.

In this context, the Nexus concept has a major advantage opposite other water management approaches on the regional level: it is multicentric and therefore can offer alternative, less sensitive pathways to deal with water-related issues in a broader development context. By addressing water challenges in a broader context as one aspect of sustainable development and a Nexus approach, can help to achieve a constructive discourse and commonly accepted solutions.

REFERENCES

Abdullaev I. and Rakhmatullaev S. (2016). River basin management in Central Asia: evidence from Isfara Basin, Fergana Valley. *Environmental Earth Sciences*, **75**, 870. https://doi.org/10.1007/s12665-016-5409-8.

Abdullaev I., Kazbekov J., Manthritilake H. and Jumaboev K. (2010). Water user groups in Central Asia: emerging form of collective action in irrigation water management. *Water Resources Management*, **24**, 1030.

Ahmadov I., Tsani S. and Aslanli K. (2011). Sovereign Wealth Funds as the Emerging Players in the Global Financial Arena: Characteristics, Risks and Governance in Sovereign Wealth Funds: New Challenges for the Caspian Countries. Public Finance Monitoring Center, Khazar University, Azerbaijan and Revenue Watch Institute NY, USA.

Alfrord D., Kamp U. and Pan K. (2015). Assessment of the role of glaciers in stream flow from the Pamir and Tien Shan mountains. Report No: ACS12128, The World Bank.

Dukhovny V. A. and de Schuetter J. L. G. (2011). Water in Central Asia: Past, Present, Future. CRC Press, Boca Raton, MA, USA.

European Bank for Reconstruction and Development (EBRD) (2004). *Transition reports*, ISBN: 1 898802 25 4.

European Commission. *Water Framework Directive (WFD 2000/60/EC)*, http://ec.europa.eu/environment/water/water-framework/index_en.html (accessed 23 September 2017).

E. Freedman and M. Neuzil (eds.) (2016). The Environmental Crises in Central Asia: From Steppes to Seas, from Deserts to Glaciers. EarthScan Series, Routledge, pp. 1–196.

Frenken K. and Food and Agriculture Organization of the United Nations (FAO) (2013). Irrigation in Central Asia in figures: AQUASTAT Survey – 2012. FAO, Rome.

Fritsch O., Adelle C. and Benson D. (2017). The EU Water Initiative at 15: origins, processes and assessment. *Water International*, **42**(4), 425–442. https://doi.org/10.1080/02508060.2017.1330816

Granit J., Jägerskog A., Lindström A., Björklund G., Bullock A., Löfgren R., de Gooijer G. and Pettigrew S. (2012). Regional options for addressing the water, energy and food nexus in Central Asia and the Aral Sea Basin. *International Journal of Water Resources Development*, **28**(3), 419–432, doi: 10.1080/07900627.2012.68430.

[17]The SPECA Working Group on Water, Energy and Environment is a subsidiary body within the governing structure of the UN Special Programme for the Economies of Central Asia.

International Water Management Institute (IWMI) (2008). *Areas of physical and economic water scarcity*, UNEP/GRID, Arendal Maps and Graphics Library.

IRENA (2016). Solar Pumping for Irrigation: Improving Livelihoods and Sustainability. The International Renewable Energy Agency, Abu Dhabi.

Kalyuzhnova Y. (2011). The National Fund of the Republic of Kazakhstan (NFRK): from accumulation to stress-test to global future. *Energy Policy*, **39**(10), 6650–6657.

Karatayev M., Hall S., Kalyuzhnova Y. and Clarke M. (2016). Renewable energy technology uptake in Kazakhstan: policy drivers and barriers in a transitional economy. *Renewable and Sustainable Energy Reviews*, **66**, 120–136.

Karimov A., Smakhtin V., Mavlonov A. and Gracheva I. (2010). Water 'banking' in Fergana valley aquifers – a solution to water allocation in the Syrdarya river basin? *Agricultural Water Management*, **97**(10), 1461–1468.

Karimov A. K., Smakhtin V., Mavlonov A. *et al.* (2015). Managed Aquifer Recharge: Potential Component of Water Management in the Syrdarya River Basin. *Journal of Hydrologic Engineering*, **20**(3). https://doi.org/10.1061/(ASCE)HE.1943-5584.0001046

Kazakhstan Green Energy-Water Conservation (2017). https://www.kzgreenenergy.com/water-conservation/ (accessed 19 September 2017).

Keskinen M., Guillaume J. H. A., Kattelus M., Porkka M., Räsänen T. A. and Varis O. (2016). The water-energy-food Nexus and the transboundary context: insights from large Asian rivers. *Water*, **8**(5). https://doi.org/10.3390/w8050193

Pala C. (2011). In Northern Aral Sea, rebound comes with a big catch. *Science*, **334**(6054), 303, doi: 10.1126/science.334.6054.303.

Pomfret R. W. (2006). The Central Asian Economies Since Independence. Princeton University Press, Princeton, NJ, USA.

Rahimon R. M. (2012). Evolution of Land Use in Pastoral Culture in Central Asia with Special Reference to Kyrgyzstan and Kazakhstan. In: Rangeland Stewardship in Central Asia, V. Squires (ed.), Springer, Dordrecht.

SCN-Lavalin. *Feasibility Study for Kambarata HPP-1*, http://www.snclavalin.com/en/projects/feasibility-study-for-kambarata-hpp-1.aspx (accessed 03 August 2017).

Sehring J. and Diebold A. (2012). From the Glaciers to the Aral Sea: Water Unites. Trescher Verlag, Berlin.

Tsani S. (2013). Natural resources, governance and institutional quality: the role of resource funds. *Resources Policy*, **38**(2), 181–195.

Tsani S., Ahmadov I. and Aslanli K. (2011). Governance, Transparency and Accountability in Sovereign Wealth Funds: Remarks on the Assessment, Rankings and Benchmarks to Date, in Sovereign Wealth Funds: New Challenges for the Caspian Countries. Public Finance Monitoring Center, Khazar University, Azerbaijan and Revenue Watch Institute NY, USA.

UNECE (2017) Reconciling Resource Uses in Transboundary Basins: Assessment of the Water-food-energy-ecosystems Nexus in the Syr Darya River Basin. Geneva, Palais des Nations CH – 1211 Geneva 10, Switzerland.

UNEP, UNDP, UNECE, OSCE, REC and NATO (2011). *Environment and Security in the Amu Darya Basin*, http://www.zaragoza.es/contenidos/medioambiente/onu/1171-eng.pdf (accessed 03 August 2017).

Wegerich K., Kazbekov J., Lautze J., Platonov A. and Yakubov M. (2012). From monocentric ideal to polycentric pragmatism in the Syr Darya: searching for second best approaches. *International Journal of Sustainable Society*, **4**(1/2), 113–130.

World Bank (2014). *Assessment Studies for Proposed Rogun Hydropower Project in Tajikistan*, http://www.worldbank.org/en/region/eca/brief/rogun-assessment-studies (accessed 03 August 2017).

ZOI (2013). Waste & Chemicals in Central Asia: A visual synthesis, https://issuu.com/zoienvironment/docs/waste-chemicals-ca-en (accessed 23 September 2017).

Assessments of the water–food–energy–ecosystems Nexus in transboundary basins: focus on lessons learned and opportunities for Central Asia

A. Lipponen[18]

United Nations Economic Commission for Europe

ABSTRACT

The European Commission and other key stakeholders engagement in Central Asia, particularly in the Syr-Daria basin, is discussed in this section. The activity was conducted by reviewing key WEFE Nexus issues, competing water uses, challenges and tradeoffs in the agriculture and energy sectors. Recognizing the need for efficient water use, essential for progressing towards the SDG 6 and 7, the main lessons learned from the past and ongoing experiences in the region are listed and discussed.

Keywords: WEFE Nexus, Central Asia, Syr-Daria

CS15.1 INTRODUCTION

This paper aims at contributing to the reflection promoted by the Joint Research Centre for development of a Position Paper for the European Commission's (EC) Directorate-General for International Cooperation and Development (DG DEVCO) with the stated main objective to "(i) frame the context for operationalizing the water–energy–food (WEF) Nexus outside EU for development cooperation support and provide recommendations for future projects and initiatives".

Conflicting objectives of different economic sectors regarding water use are commonly a source of tension. Water security is linked to other resource securities – notably food and energy – and environmental security, and these interlinkages need to be taken into account in planning and management for overall sustainability. The interconnectedness of sectorial development goals is evident in Agenda 2030 for Sustainable Development (ICSU, 2017), underlining the need to identify synergies and minimize trade-offs in planning. A "nexus approach" to managing interlinked resources is

[18]The author is responsible for the area of work on the water–food–energy–ecosystems Nexus under the UNECE Water Convention. She coordinated the development of the methodology, the basin assessments and serviced the related intergovernmental process. The views expressed in this article are those of the author and do not necessarily represent the views of the United Nations or its member states.

increasingly recognized as a way to enhance water, energy, and food security by increasing efficiency, reducing trade-offs, building synergies and improving governance while protecting ecosystems.

Resource security is too commonly inferred to be best achieved through national means when transboundary cooperation and intersectorial coordination at that level may be more effective strategies to that end. Support, investment, and partnerships that take into account interconnectedness of resources and sectors are expected to be more effective and more sustainable.

The experience synthesized in the present paper is drawn from the five assessments of the water–food–energy–ecosystems nexus in transboundary river basins[19] under the Convention on the Protection and Use of Transboundary Watercourses and International Lakes (signed in Helsinki in 1992) as part of the Convention's Programme of Work 2013–2015 and 2016–2018 (UNECE, 2012, 2015a). The Nexus approach demonstrated by these assessments invites to consider the effects on resources broadly, not just through for example, water uses and discharges, but how sectorial policies directly and indirectly influence the dynamics between the resources in focus (water, energy, and land/food resources as well as ecosystems) as well as the respective management spheres and where improved sustainability can be achieved through joint action. The basins that were assessed are located in the Caucasus, Central Asia, and Southern Europe. These participatory assessments were prepared in close cooperation with the concerned Ministries of the riparian countries sharing the respective transboundary basins. In practice, they involved a study of intersectorial links, trade-offs and benefits in managing water, energy, land, and environmental resources, and an accompanying intersectorial transboundary dialogue about the nexus issues and possible solutions, informed by the study (analysis).

CS15.2 STATUS OF APPLYING THE NEXUS CONCEPT AT THE TRANSBOUNDARY LEVEL AND ASSESSMENT APPROACHES

Giving a review of how the Nexus concept is applied in transboundary contexts is outside the scope of this short paper. With somewhat different definitions and scoping, various initiatives involving intersectorial or concurrent multisectorial analysis have been undertaken to study complex interlinkages covering energy, water, and food or agriculture or at least some of these sectors (or resources).[20] The paper focuses on the practical experience from the transboundary participatory assessment made under the Water Convention.

CS15.3 THE METHODOLOGY

For the purpose of the assessments, a devoted methodology was designed with the support of expertise representing different sectors, piloted and subsequently applied in river basins and an aquifer selected on the basis of expressions of interest from riparian countries or from transboundary cooperation organizations. The basins differ markedly in size, geography, level of cooperation, characteristics of the economic sectors, and by other features. The methodology involves six steps which along with the inputs and outputs of each step is shown in graphical form in Figure CS15.1. For details of the methodology, UNECE (2015b), de Strasser *et al.* (2016) and, for the governance aspects in particular, UNECE (2017a) can be referred to. The tasks specified for each step are either carried out by analysts in desk studies and

[19]Information about the basin assessments is available at http://www.unece.org/env/water/nexus.html

[20]A number of tools and initiatives were presented at the global stocktaking Workshop on Assessments of the Water-Food-Energy-Ecosystems Nexus and Response Measures in Transboundary Basins, held at the Palais des Nations, Geneva, 6–7 December 2016. The report of the workshop can be referred to for more information on these (UNECE, 2017a).

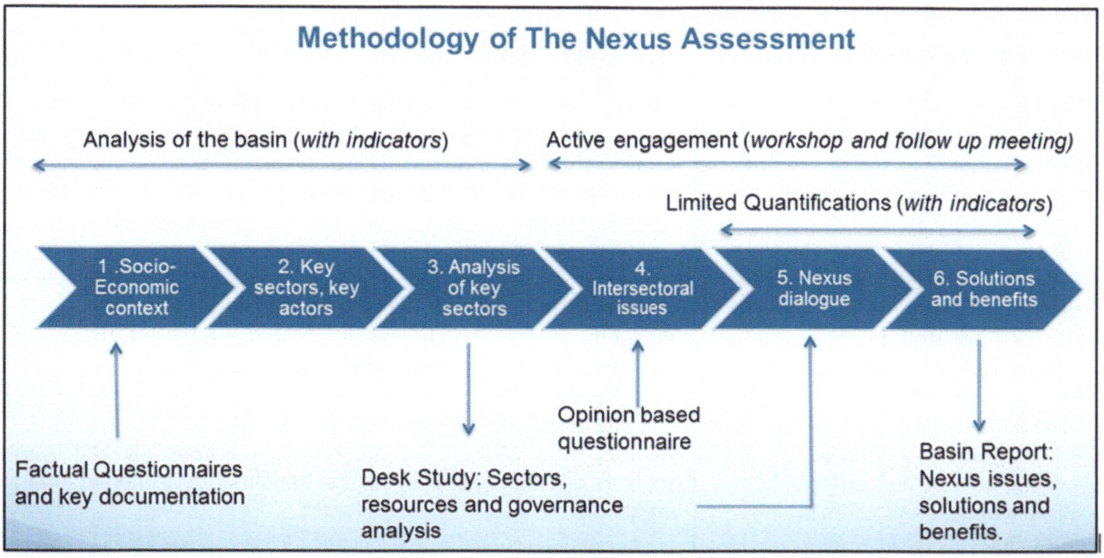

Figure CS15.1 Different stages of and inputs to the nexus assessment methodology developed under the Helsinki Water Convention for application in transboundary basins (de Strasser *et al.*, 2016).

analysis (involving quantification with fit-for-purpose tools) or by the authorities and stakeholders during workshops and consultations. The different steps are described in Table CS15.1.

The Task Force on the water–food–energy–ecosystems Nexus was established by the Meeting of the Parties to the Water Convention (UNECE, 2012) to guide the work on the nexus assessments and to provide oversight. Because of the key role that the Governments of the countries sharing the basins assessed had in steering the process, the official participation in the process and the various consultations ensured that the assessments reflect the challenges perceived as relevant by the Governments.

CS15.4 KEY FINDINGS

CS15.4.1 Lessons from nexus assessments in transboundary basins

Improvements to management of interlinked water, energy, land, and ecosystem resources are not only about financing and technology, but also about mind-sets and behaviours, ways of planning and implementing interventions as well as governance. For identifying the locally relevant nexus issues and for seeking commitment to address them, the following can be highlighted from what has emerged as important[21]:

- A participatory process that seeks to effectively engage all the concerned countries (the basin sharing countries in this case), and all the main sectors and key stakeholders in the process has a particular value for ownership of the conclusions, for integrating local knowledge and for building capacity.
- The need to build on complementary institutional frameworks: Ensuring involvement of the key sectors is crucial and so is selection of the institutional platform(s) used for conducting the assessment and the dialogue. Water management is an important connecting element, but engaging

[21]For a more extensive discussion about lessons learned, UNECE (2015b) and de Strasser et al. (2016) can be referred to.

Table CS15.1 The six steps of the nexus assessment methodology.

Step and the Modality	Inputs and Outputs
Step 1 A desk study and an analysis of existing data	The current and if possible projected needs of the population in the basin as well as the national needs that rely on the basin are identified. This is done by looking for example, at the water, energy and land resources as well as their uses. This develops an understanding of the basins socio-economic context, its resource base and the governance context.
Step 2 A desk study and an analysis of existing data	The identified needs are associated with sectors and institutions. Hence, the key sectors and stakeholders are identified, also to contribute to the process.
Step 3 A desk study and analysis of existing data as well as input from officials and stakeholders during a workshop	The key sectors are analysed applying the drivers-pressures-state-impacts-response framework. The analysis is further refined in the first transboundary intersectorial workshop.
Step 4 Input from officials and stakeholders during a workshop	In the first workshop, officials and other key stakeholders identify and detail issues between sectors while considering the sectorial strategies and development plans and linkages to other sectors. This includes material presented by experts, officials, or stakeholders.
Step 5 Input from stakeholders during the first workshop as well as through a desk study and analysis of additional data	Nexus diagrams about the main interlinkages are reviewed, complemented, and validated collectively with the stakeholders (including through group works). It includes the nexus components water, food, energy, ecosystems, and the significant linkages identified. Further analysis by experts using additional data refines and details the linkages.
Step 6 With assistance of stakeholders during a workshop as well as by analysis through a desk study and analysis of additional data	Possible solutions to the most pressing intersectorial issues are identified, such as, land-use management, cooperation agreements, policy solutions, infrastructure projects, or economic instruments. These are discussed in the later workshops.[22]

with the energy sector's or other economic frameworks is necessary for promoting awareness and revisiting practices where sectorial policy decisions are taken.

Regarding lessons, Central Asia merits particular attention in this paper, considering the European Commission's engagement in the region.[23] Nevertheless, the findings from the Water Convention's

[22]Based on practical experience from the application of the methodology, at least two workshops should be organized in the assessment process, the latter providing for discussion about the findings, either as preliminary or further developed.

[23]A regional EC Nexus Dialogue programme is carried out in Central Asia with the support of the EC by https://www.nexus-dialogue-programme.eu/about/nexus-regional-dialogue-programme/

nexus assessment are more generally relevant, demonstrated by, for example, invitation to experience sharing for other regional EC Nexus Dialogues programme supported by the EC and the Deutsche Gesellschaft für Internationale Zusammenarbeit (GIZ) in the Niger Basin and in Latin America.

CS15.4.2 Syr Darya Basin

The current situation in the Syr Darya Basin is heavily impacting on the development in the riparian countries (Kazakhstan, Kyrgyzstan, Uzbekistan, and Tajikistan) and the basin. Examples of issues that will likely affect the interlinkages with regard to water quality and water quantity are: the lack of a functioning energy market, the fact that water is not valued and priorities of development are different between countries, that there is neither an effective pollution control nor functioning incentives for improving resource efficiency. Important trends are population growth with increased demand for food, water, and energy, and increased pressure on natural resources and the environment. Key uncertainties that will affect the impact of these trends are the development of regional co-operation, geopolitics, climate change impact, and emergencies (e.g., droughts). The main nexus interlinkages in the Syr Darya Basin, identified jointly with the various ministries and other key stakeholders in the assessment,[24] are presented in Figure CS15.2.

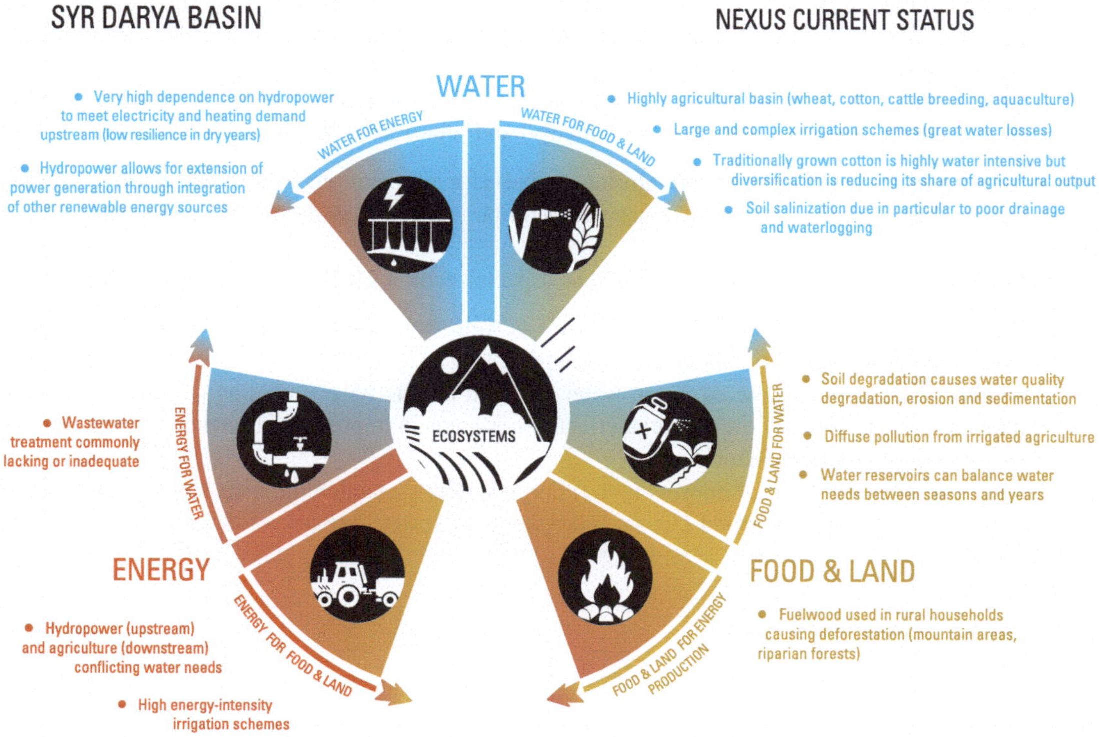

Figure CS15.2 Main nexus interlinkages in the Syr Darya Basin (UNECE, 2015b).

[24]It should be noted that Uzbekistan does not associate itself with the nexus assessment of the Syr Darya.

Natural resources in the Syr Darya Basin could be developed and used more effectively and with less impact on the environment considering the basin more holistically and capitalizing on the complementarities between the countries. The results of the nexus assessment suggest that adoption of the Nexus approach has the potential to improve resource use efficiency and security in the riparian countries. In contrast to national approaches presently employed, cooperation involving all the countries and sectors has significant potential to optimize the use of resources in the basin.

At the same time, applying certain solutions at the country level – including, among others, improvement of efficiency in water and energy use, as well as well-targeted economic and policy instruments – can help gradually build more favourable conditions for transboundary cooperation.

The solutions proposed in the assessment can be summarized by the following categories (UNECE, 2017b):

- Improving energy efficiency, reducing dependency on water for energy (diversification of sources), and rationalizing water use (especially in agriculture);
- Developing a well-functioning regional energy market and exploring opportunities for energy–water exchanges between countries, the development of alternative energy sources, and improving overall energy efficiency;
- Lowering barriers to trading food and agricultural goods, thus promoting their more cost-, water-, and energy-efficient production and exchange within the region;
- Developing mechanisms to incorporate wider impacts in sector-based policy development, and improving inter-sectorial coordination at the basin level by increasing representation of and consultation with the relevant ministries;
- Improving basin-wide monitoring, data verification and exchange, and knowledge-sharing, including joint monitoring (e.g., of water flows and quality) and joint forecasting.

The Syr Darya assessment provides a good information basis for interventions by different actors. A number of more technical and detailed studies have been carried out, referred to in the Syr Darya report (UNECE, 2017b).

Only to mention a few, the following emerged as aspects needing further study in Central Asia:

- Governance across sectors at the national level, as well as the application and outlook for selected intersectorial processes (e.g., SEA) in the countries.
- Available water resources, saved water, water and land productivity, including a critical descriptive and quantitative analysis of the potential effect of measures in improving efficiency of irrigation and improvement of land reclamation.
- Cooperation opportunities for improving the energy system sustainability, including funding options and mechanisms.[25] The feasibility, costs and benefits of selected nexus solutions prioritized by the countries.[26]

[25]The four-country energy system model developed by KTH with an open-source tool for the Syr Darya nexus assessment could be used to assess impacts of development, including investments. The precision of related insights could be improved by validation of generation infrastructure information (including for planned projects), as well as detailing of data on electricity trade and capacities, on fuel prices, and transmission and distribution losses.

[26]For a comprehensive list of the proposed nexus solutions UNECE (2017b) can be referred to.

CS15.5 DISCUSSION

Due to the interlinkages between the SDGs on food security (SDG 2), water and sanitation (SDG 6) and sustainable energy (SDG 7), well-targeted measures in economic sectors can have benefits beyond the respective SDGs: As a case in point, increased use of renewable sources of energy and improved energy efficiency will, in most cases, reinforce targets related to water access, scarcity and management by lowering water demands (ICSU, 2017). While wind power is more expensive than hydropower, it could be pursued at selected sites to diversify domestic energy supply and enhance energy security. Indicative calculations made by the Royal Institute of Technology in Stockholm (KTH) for the Syr Darya assessment demonstrate that a combined wind and hydropower programme in Kyrgyzstan and Tajikistan could bring significant water savings (UNECE, 2017b). Agriculture being the main consumptive use of water, more sustainable agricultural practices are key to ensuring meeting a number of water targets under SDG 6, helping to progress towards SDG 2. In addition to introducing and promoting energy and water-efficient technologies, development of economic instruments is another means of rationalizing use of these resources. For example, OECD has outlined possibilities for applying such instruments in Kyrgyzstan (OECD, 2014).

Central Asia is progressing in gradually introducing principles of international law into national laws and policies, as well as standards and good practices. Related to the environment, the progress is documented in the Environmental Performance Reviews.[27] A recent study shows that, even in introducing the key aspects of IWRM, a lot yet remains to be done by Central Asian countries to make basin level management and participation of different stakeholders well-functioning (UNECE & OECD, 2016). The Nexus approach aspires to take integration across sectors even further. Improving policy coherence is a long-term effort, and many EU and UNECE instruments have been developed that are consistent and complementary. They therefore provide a useful reference.

Evidence-based decision-making should be encouraged and supported by improving availability and sharing of relevant information, and facilitating its uptake in policy. Information needs for (quantitative) analysis of the nexus issues across sectors are however high and constraints in availability of or access to data may turn out limiting. For example, availability of up-to-date data on the status of water resources and ecosystems is in many cases not good. Already at the national level, information across sectors is not necessarily easily available. At the transboundary level, information exchange requires improvement, more regularity, continuity, transparency, and structure (UNECE, 2011).

The key is to have solid studies informing policy development and decision making, involving local institutions and local experts, both to transfer knowledge and to integrate in-depth local understanding. However, the conduciveness of official frameworks and processes to consideration of studies varies. The Nexus Dialogue project can be expected to contribute to building capacity and awareness-raising regarding the nexus issues.

The Water Convention's nexus assessments from the Caucasus, the Western Balkans, and Southern Europe highlight a number of potentially valuable experiences that are relevant also for Central Asia: Co-optimizing flow regulation (Drina River Basin; UNECE, 2017c), improving water-use efficiency in agriculture (the Isonzo/Soča Basin) and improving access to modern energy to improve livelihoods with co-benefits for improving the status of water resources and ecosystems through reduced erosion (Alazani/Ganykh River Basin) (UNECE, 2015b).

With the recent positive developments in the relations between the Central Asian countries, the outlook for major projects of transboundary and/or regional significance has also improved. Nevertheless, water

[27] Available from: http://www.unece.org/environmental-policy/environmental-performance-reviews/enveprpublications.html

remains a subject of conflicting interests. It is positive, though, that the 'Joint Communiqué: European Union – Central Asia Foreign Ministers' Meeting 'EU and Central Asia: Working for a Safer and More Prosperous Future Together' (Samarkand, 10 November 2017)[28] recognizes that "Continued and expanded dialogue and cooperation contributes to the efficient use of water and energy resources, the protection of the environment and for addressing climate change as well as other cross-border issues".

CS15.6 CONCLUSIONS AND RECOMMENDATIONS

The EC engagement and support in Central Asia has the potential to assist the region in moving towards more sustainable development of water and related resources. However, a one-off project or even investment programme may have a limited effect in the long term, unless accompanied by assistance to the countries to further develop their legal basis, policies and procedures that can provide for more coordinated and consultative planning and resource management, including for investments, in the long term. In particular, strengthening institutional frameworks that can provide for dialogue, coordination and assessment of impacts would be valuable.

The Syr Darya nexus assessment called for "further analytical, stakeholder engagement and planning work to identify precise governance reforms, policy measures and investment opportunities to address all the challenges and seize any corresponding opportunities". It points at a number of beneficial actions that any further work on the nexus issues in Central Asia can build on. Supporting diversification of energy sources, energy trade and improved energy efficiency, and furthering the on-going transformation of agriculture involving improved efficiency of water use, crop switching, and land reform, are examples of areas identified where investment and other measures reduce the demands and dependencies behind the current main trade-offs in the nexus. Facilitating trade of agricultural products and addressing the current regulatory and procedural barriers to trade can support further transformation of the agriculture sector.[29]

There are still significant development needs in the region and many investment projects are therefore still oriented towards expanding supply when some such investments could perhaps be avoided by managing demand. Currently, a business case is still commonly lacking for investing into e.g. efficiency in the use of water and energy. For energy efficiency, identification of obstacles to investment in a recent UNECE report is instructive (UNECE, 2017d). Improving efficiency is crucial for progressing towards both SDG 6 and SDG 7. Support can build on and use experiences from efforts towards efficient resource use: Uzbekistan and Kazakhstan have reported investments into water-efficient irrigation (UNECE, 2017b) and in the Kyrgyz Republic, effective irrigation technologies and the outlook for their local application have been investigated (MAM, 2015).

A more enabling environment for investments that considers impacts on the environment and across sectors can be developed. The international financing institutions' procedures and conditions as well as the scope of their investment and related technical assistance programmes also represent relevant factors in moving towards a more sustainable development approach. The EC's cooperation with international financial institutions can help integrate the nexus perspective better into financing. An event organized

[28]The Foreign Ministers of the Republic of Kazakhstan, Kyrgyz Republic, Republic of Tajikistan, Turkmenistan, Republic of Uzbekistan, the European Union's (EU) High Representative for Foreign Affairs and Security Policy and Vice-President of the European Commission, and the European Commissioner for International Cooperation and Development met on 10 November 2017 in Samarkand, Uzbekistan, for the 13th EU-Central Asia Ministerial Meeting.

[29]Studies of regulatory and procedural barriers in Kazakhstan, Kyrgyzstan, and Tajikistan by UNECE Trade Programme:http://www.unece.org/trade/publications.html

by the EC through the European Union Technical Assistance Facility (TAF) and UNECE's Sustainable Energy Division hosted on the occasion of the 8th International Forum on Energy for Sustainable Development and Ministerial Conference in June 2017 in Astana, Kazakhstan opened an initial pipeline for identification of renewable energy investment projects. Extending investment discussions to broader groups of stakeholders can lead to wider support and better identification of 'nexus' opportunities, including synergies between initiatives. Earlier in a policy cycle, well before implementation of measures (including investments), at the national level, the European Union Water Initiative's National Policy Dialogue provides in four Central Asian countries a platform for inter-ministry coordination in developing strategic documents, policies and pieces of legislation (UNECE & OECD, 2016). As an example of institutionalizing a consultative process to assess the effects of policies is Kazakhstan's pilot application of the Strategic Environmental Assessment to the future development options of the national energy system.

Many impacts in the nexus are transmitted through water. As the region's main rivers are transboundary, transboundary impacts (and benefits) of major development projects need to be evaluated. Requirements of financing institutions are influential but the Water Convention and other UNECE instruments are also of support. Some major gaps remain: For example, only with an exception, transboundary procedures related to the Environmental Impact Assessment (EIA) are missing in national legislations of Central Asian countries. Updating of the guidelines for applying the Convention on the EIA in a Transboundary Context in Central Asia, once completed (on-going in 2018) will be a resource for the region, although legal reforms would be needed for a greater effect.

Gradually adopting more cooperative solutions, including planning coordination between the countries, would reduce pressure on shared resources. Developing the institutional frameworks for cooperation is important to that end. It complicates addressing some of the most problematic nexus issues in Central Asia that the frameworks for regional water cooperation do not include the energy sector. At the same time, like the agriculture sector, the energy sector is gradually transforming and modernizing itself – reducing emissions, exploring the potential of renewable energy sources beyond hydropower etc. Decisions affecting the energy mix or about interconnections could potentially offer at least partial solutions to water challenges that are difficult to address by water management alone.[30]

REFERENCES

de Strasser L., Lipponen A., Howells M., Stec S. and Bréthaut C. (2016). A methodology to assess the water energy food ecosystems nexus in transboundary River Basins. *Water*, **8**(2), 59.

ICSU (2017). A Guide to SDG Interactions: From Science to Implementation. International Council for Science (ICSU), Paris.

MAM (2015). Modern Irrigation Technologies: Recommendations on Implementation in Kyrgyzstan. Ministry of Agriculture and Melioration of the Kyrgyz Republic (MAM), Department of Water Management and Melioration, Bishkek, November 2015. Available from: http://www.unece.org/fileadmin/DAM/env/water/meetings/NPD_meetings/Publications/2015/Study_of_suitable_irrigation_technologies_for_Kyrgyzstan_ENG.pdf.

OECD (2014). Reforming Economic Instruments for Water Resources Management in Kyrgyzstan: Proposed Options for Reform. Draft Final report (April 2014).

UNECE (2011). Second Assessment of Transboundary Rivers, Lakes and Groundwaters. United Nations, New York and Geneva.

[30]Most of the documents are available either from http://www.unece.org/env/water/publications/pub.html or at the pages of the respective meetings at http://www.unece.org/environmental-policy/conventions/water/envwatermeetings.html#/

UNECE (2012). Report of the Meeting of the Parties on its seventh session (Rome, 28–30 November 2012), Addendum, Programme of work for 2013–2015 (ECE/MP.WAT/37/Add.1.).

UNECE (2015a). Report of the Meeting of the Parties on its seventh session (Budapest 2015), Addendum, Programme of work for 2016–2018.

UNECE (2015b). Reconciling Resource Uses in Transboundary Basins: Assessment of the Water-Food-Energy-Ecosystems Nexus. United Nations, New York and Geneva.

UNECE (2017a). Revised Governance Methodology for Assessing the Water-Food-Energy-Ecosystems Nexus. Task Force on the Water-Food-Energy-Ecosystems Nexus, Fifth meeting (Geneva, Switzerland, 18 October 2017).

UNECE (2017b). Reconciling Resource Uses in Transboundary Basins: Assessment of the Water-Food-Energy-Ecosystems Nexus in the Syr Darya River Basin. United Nations, New York and Geneva.

UNECE (2017c). Assessment of the Water-Food-Energy-Ecosystem Nexus and Benefits of Transboundary Cooperation in the Drina River Basin. United Nations, New York and Geneva.

UNECE (2017d). Overcoming Barriers to Investing in Energy Efficiency. United Nations, New York and Geneva.

UNECE and OECD (2016). Implementation of the Basin Management Principle in Eastern Europe, the Caucasus and Central Asia. European Union Water Initiative National Policy Dialogues Progress Report 2016 (May 2016).

5

Conclusions and recommendations

When designing a Nexus assessment, a number of aspects need to be considered, including data availability, temporal and spatial scale, stakeholder involvement, impacts on ecosystems at the local, regional, and global level, socio-economic inequalities, and a need for interdisciplinary expertise. A number of Nexus tools and methodologies are currently available but in the majority of the cases these are tailored for specific quantitative applications and cannot be easily transferred to more general Nexus approach. For this reason, applying a qualitative representation of interconnections is as important as the quantitative assessments.

Assessing synergies and trade-offs between water, energy, food security and ecosystems allows for investigating interdependencies between different Nexus dimensions, and the interrelations between the Nexus and the SDGs. This can include qualitative and quantitative analyses of economic dynamics, biodiversity, sustainable development challenges and climate change impacts. Integration of current models and methods can allow for scenarios analyses of the complex mix of components such as sectorial and future development strategies, decision-making processes in order to identify possible trade-offs and synergies for formulating information-based policy.

Locations where resource competition and issues can represent a Nexus hotspot for small-scale testing. Public participation and direct consultations with stakeholders provide demand-driven inputs and encourage buy-in for the relevance and the uptake of a Nexus approach. The stakeholder engagement should be wide and include ministries, regulatory bodies, local administrations, services, educational/research institutions and the private sector who should be actively engaged in the process of design, setup and development of a Nexus strategy and incorporation into a planning process. It is recommended to use existing platforms for Nexus dialogues and implementation, rather than creating a new platform which can be difficult to attribute sufficient legitimacy. Such platforms could include regional economic commissions, river basin organizations or inter-ministerial committees and other national frameworks such as SEAs.

5.1 FROM THE NEXUS DIALOGUES WORKSHOP

The results from the Nexus Dialogues Workshop of 2018 assemble conclusions over three sections which can be described as follows:

5.1.1 Applying sustainable technological approaches and solutions

The sub-themes proposed were:

- The role of off-grid solutions for example, for water pumping or desalination in breaking the fossil-fuel economic dependency, including as applied to food security;
- From waste to resources: opportunities to turn wastewater and agricultural waste into nutrients and energy, and associated risks;
- The levels at which to apply a Nexus approach to competency and capacity building to promote 'Nexus thinking' and take action across sectors, countries, and areas of responsibility.

Case studies indicate that the choice of hybrid technologies, for instance combining several sources, can fit local conditions and improve sustainability of energy and water resources for agricultural and in particular irrigation purpose. Off-grid solutions to satisfy energy and water needs were also found to be cost effective at local scale (e.g., farm scale), even though there are still technological challenges with providing a steady supply and also storage for intermittent use of off-grid solutions. Furthermore, renewable energy-based off-grid solutions have a role in supplementing fossil fuel reliance rather than representing a generic alternative to them. They have the potential to reduce the amount of water used as well as increase food security and water availability in remote areas. The optimization of national and regional grids and plants facilitates the deployment of additional energy capacities. Turning waste (from e.g., agriculture activities) into a resource can traverse sectors; crop residues or other wet agricultural waste like manure can substitute wood charcoal for energy. At community level, implementing such solution requires capacity building and promotion of the Nexus solution which must be practical and have social acceptance.

5.1.2 Data, Nexus tools, and models

The sub-themes proposed were:

- The challenges to integrate or link quantitative models focusing on multiple Nexus pillars and dimensions, that is, water, energy, land, climate change, the environment, and others;
- The trade-off between the level of sophistication of Nexus Decision Support Tools and the level of integration across scales for effective modelling and meaningful decision making;
- The opportunities and risks of an open-source approach to Nexus tools and data sharing as opposed to commercial packages and solutions.

In general, Nexus experts found that lack of data availability is generally a strong limitation for appropriate Nexus assessments. Consequently, there is a general need to improve data collection across the different sectors and scales, and to develop a general strategy oriented towards an open data policy. Some potential alternative data sources were also discussed, as for the case of remote sensing and field surveys, but the need for a more systematic data management system was strongly highlighed. Resolving this issue was identified as a key milestone to supporting decision making in the fields of food security, poverty reduction, sustainable basin management, and inclusive development. While acknowledging the general open data issues, the experts agreed that open-source models and open-access platforms combining spatial data have been well received in low- and middle-income countries where license costs and governmental authorization may be difficult to obtain.

5.1.3 Governance, finance, institutions, and cooperation frameworks

The sub-themes proposed were:

- The requirements for investment-ready Nexus projects with potentially conflicting objectives, i.e., financial and economic vs. social and environmental in a portfolio approach to finance;
- The river basin scale as the most appropriate management framework for approaching and implementing the Nexus;
- The advantages of promoting a Nexus approach within the ongoing debates and actions surrounding climate change mitigation and adaptation;
- The operational aspects of the Nexus as a platform to guide sustainability efforts underpinned by the SDGs;
- The definition, utility, and use of a Nexus project toolkit for inclusive green growth and sustainable development.

A key conclusion is that new innovative financing mechanisms are needed. In fact, when analysing and looking at investment decisions, attention is usually focused on two parameters: return on investment or simply profitability (i.e., measured investment efficiency), and risk (including in terms of impact on society and the environment). Risk is seen as particularly important because Nexus projects include social elements that are not commercially viable.

From a governance point of view, the Nexus approach does not require new institutional structures. What is needed is comprehensive collaboration between entities such as working networks or platforms supported by multi-sector policies, protocols, and procedures. Climate and environmental actions and the Nexus approach can work hand in hand; the success in one is intricately linked to success in the other. The Nexus approach requires multi-sectorial thinking and the raising of awareness and can deal with the impacts of climate change and facilitate the development of national and transboundary Basin Adaptation Plans.

5.2 LESSONS LEARNED

In general, current sector models are largely limited to individual sectors and hence not entirely fit for untangling the complex interlinkages of the Nexus, especially beyond the local scale at which most models are being applied. Computer modelling can be complex and accessing data, unless open source, can be costly. However, commercial packages can offer longer-term support compared to free open-source software and tools which often do not have access to skills and finance to guarantee the adequate level of quality. There will be concerns on data-sharing, such as for national security within national governments and regional entities such as river basin organizations, power pools or transportation frameworks; and these can create bottlenecks or block data exchange. The following section forms the basis of the "EC Position Paper on Water, Energy, Food and Ecosystems (WEFE) Nexus and Sustainable Development goals (SDGs)". Organized around the conclusions and recommendations of the different thematics with the objective of summarizing the experiences coming from the Nexus experts and the discussions held during the workshop, the section summarizes the identified knowledge gaps and implementation challenges, as well as further research areas and practical actions to be implemented:

- Accessing finance for the Nexus is underpinned by a project's bankability and risk management plays an important role in this. Projects that potentially make up a SDG-oriented programme for improved access to affordable, reliable, and modern energy services, increased agricultural productivity and universal and equitable access to safe and affordable drinking water, vary in terms of risk profile and commercial potential.

- Nexus programme can have increased commercial potential with public–private partnerships implemented under a blended finance approach. For example, irrigation projects accessing a mix of grant funding and concessional debt from development finance institutions could meet capital expenditure needs and apply a nominal levy to fund operational and recovery costs of a scheme. The positive economic and social value of the irrigation schemes enhances the possibility of securing concessional debt finance for the capital cost of the project.
- Investment banks and financial markets are increasingly looking socially responsible, ethical and green investments and adjusting risk management to incorporate green loans, and supporting businesses and industries aiming for a positive impact on the environment and society. This can include green bonds for funding for climate-friendly projects clean renewable energies, sustainable and bio-diverse land and water use, on which Nexus initiatives can intuitively piggyback. The objective being to make Nexus solutions affordable until they outcompete single-sector interventions.
- Water is seen as a main driver in implementing a WEFE Nexus approach, being at the centre of food security, energy generation, economic development, ecosystems services and climate change. The River Basin has been promoted by the water and environment sectors as the most appropriate Nexus management unit with basin plans aligning with economic development and planning objectives in other sectors.
- Cities are where local and global resource constraints meet, and attempts to satisfy the resource demands of growing urban areas and consumers' lifestyles has meant looking ever further afield for supplies including at the regional and global level. Resource challenges are further exacerbated by the expanding urban populations and inefficient infrastructure systems of developing cities in several regions of the world, making the urban environment an emerging context for applying a Nexus approach.
- Connecting basins with urban context will require shaping forms of policy and governance which stretch beyond the boundary of the city and adaptation approaches to act on climate change more effectively.
- Compared to other regional water management approaches, the Nexus concept offers alternative pathways to deal with water-related issues in a broader development context. In Central Asia, the notion of the Nexus has seen an evolution in the role played by water in the past three decades. In the 1990s–2000s, the legacy of the Soviet era's water allocation principles led to integrated approaches and regional institutions even prior to the rise to prominence of the Nexus concept, however, under the paradigm that water management is exclusively a technical issue. Starting from the 2000s until recently, the region had to seek new arrangements and types of agreements for tackling interplaying water, energy, and food issues as water cooperation principles were becoming increasingly contested. With stark upstream–downstream differences in domestic water requirements for hydropower vs. agriculture, water had become an economic and political issue. Since 2015, there has been a search for pragmatic and effective solutions to otherwise difficult cooperation dialogues that focus on water as a security issue for all WEFE Nexus sectors.
- In the Latin America and Caribbean region, the Nexus approach addresses the lack of good quality information, the great diversity of the region, and generally weak governance. Some of the interconnections that should be prioritized in the region are irrigation modernization and aquifer management; biofuels; hydropower, petroleum and mining extraction; and water supply and sanitation services in cities. This Nexus planning must be adaptive, multi-scalar, multi-temporal,

inter-sectorial and based on a broad consensus by all political forces and must address social inequity. Even within Development Finance Institutions (DFIs) investment, there is a need to avoid 'cherry picking' of projects and avoid projects discriminating on the basis of the best outcome for only one sector. A Nexus approach can promote integrated investments in WEFE infrastructure which can also include financing of health, education and WASH.

- From a regulatory framework point of view, the river basin approach does not always encapsulate management challenges related to groundwater resources. Energy, and power pools in particular, tends to be regional or national in its organization as a sector and not tied to river basins. A water-specific territorial unit, on the other hand, quickly becomes preferred to the ones proper of energy or food in areas characterized by an endemic water scarcity. This is because most impacts of an uncoordinated approach to resource management are mainly felt through water in the Nexus.
- With the growing importance of the mining industry in many developing countries, there is a need to better understand the Nexus in the context of mining as well. Mining issues often come down to water quality and quantity issues for production and on ecosystems, competition for access to land, and access to reliable and cost-effective energy.

5.3 RECOMMENDATIONS

One of the most complex activity involved in the practical application of the Nexus management of complex water systems is related to the analysis of the different components and the combination of the information related to the dimensions of the Nexus itself. On the one hand, quantitative analysis through modelling is necessary to disentangle the specific dynamics of a physical, mathematical, or engineering process. On the other hand, all these different aspects need to be put in relation and coordinated in order to analyse and represent the Nexus system as a whole. In this respect, earth system models or integrated assessment models have their place in scientific research (e.g., for exploration, theory development and issue anticipation or tabling), political strategy testing and agenda setting. Going beyond the types of challenges proper of model integration, without developing overly complicating integrated models, seems to suggest the need to go beyond models. Being holistic in the way of thinking does not necessarily imply being holistic in coding models. On the contrary, more investment could go in focusing on the conceptual connection of problems, while favouring single-question models that do address these questions as simply as possible.

The most common practical solution to model integration that keeps domains separate and, consequently, inherently robust is represented by multiple possibilities of soft-linking of specific sectorial models. A key to the Nexus approach is to identify the points at which the resource systems interact and to establish appropriate data exchanges between the modules (e.g., water requirements in the land-use and energy systems but also energy needs for water supply and land use, and land requirements for energy and water infrastructure). The output from one module forms the input for the other models. This type of the multi-model framework is the most common way of dealing with complex and multiple systems. This solution, however, brings to the analysis a certain degree of simplification that needs to be accounted for in the overall Nexus systemic analysis. Given its relative simplicity and transparency, this kind of modelling framework is expected to be more widely applied in the WEFE Nexus analysis, in combination of qualitative assessment and uncertainty estimation.

This highlights the need for connecting multiple sectors, analysis, and management at different scales. The national scale may be appropriated for target setting as being the scale for the SDGs whilst municipality or watershed scales may drive most of the changes. Cities can be also considered as an appropriate scale because this is where local and global resource constrains meet.[1] Resource challenges are further exacerbated by the fast-expanding growth rates and inefficient infrastructure systems of developing cities in several regions of the world. It follows that the urban environment is an emerging context for applying a Nexus approach.

The definition, utility, and use of a Nexus project toolkit for inclusive green growth and sustainable development could include quantitative analyses of the benefits of the approach in the individual sectors, implementation guidelines, and good experiences. The upscaling of pilot Nexus projects to larger scale programmes should be part of the toolkit. This should also cover processes to identify up-scalable project elements, steps to be followed when institutionalizing the Nexus processes and dialogues, and finally pre-conditions to upscaling in terms of identifying financial and human resources.

Areas of further research and other practical actions:

- A project's financial viability is based on strengthening a variety of factors and an integrated risk can take into account: the ownership structure; the communities where the project is located; the funding requirements and the project's ability to service the debt; technology, capacity of the infrastructure; environmental analysis; market analysis and contractual and institutional arrangements. The risk management system so created can adequately integrate the complexity of water, energy and food projects, including the risk-sharing protocol among the blend of financiers required by the different risk elements.
- Regulation can help define the right criteria for Nexus projects, including sustainability criteria, and thus gradually elevate average project quality. But for beneficiary countries to move away from donor-driven agendas, this requires them to have their own (national or regional) agenda and policies in place as well as good regulatory frameworks for projects that are not just 'bankable'. For example, terms of reference that promote by-products of waste management to include recycling, re-use, waste-to-energy conversion and local market possibilities (e.g., wastewater treatment facilities generating biogas). The push for developing and applying such directives should come from comparing the cost of simple waste management to integrated management strategies and estimating financial returns based on reduced capital costs of 'industrial symbiosis'. The use of big data and social media would also help to create the demand for Nexus solutions and Nexus projects would then be possible to aggregate to build scale.
- If not the river basin, the appropriate scale of planning and management for the Nexus seems to be the one where most economic, sectorial interests congregate. In many instances, this would coincide with the sub-basin level for large rivers such as the Nile, the Amazon, the Niger, etc. but it will need to be determined on a case-by-case basis. In some cases, within regions like SADC, demand for water and energy resources is concentrated in parts of the region that are not the ones endowed with abundant water resources that could sustainably deliver those inputs. In some other cases like LAC, hydropower potential across downstream stretches of tributaries may have to be assessed for trade-offs with upstream land-use change from large-scale export-driven agriculture compounded by climate change.
- A multiple hierarchy of scales is needed anyway. The national scale may be appropriate for target-setting whilst the farm or municipality or watershed scale may drive the most change. The landscape scale is important for uniting different needs including those of ecosystems whilst food security is affected by global issues such as export prices and global energy demand. As an example, the regional scale is particularly important for SADC where watercourses and electricity grids are already shared among countries. This is the context where a regional Nexus operational

[1]Source http://www2.giz.de/wbf/4tDx9kw63gma/UrbanNEXUS_Publication_ICLEI-GIZ_2014_kl.pdf

framework is being developed to drive and guide sectorial collaboration in implementing priority investment projects. This is with a view to anchoring the Nexus in the sustainable infrastructure development and industrialization agenda.

- The implementation of the SDGs, and SDG reporting by UN Member States, must be seen as an opportunity to promote a Nexus approach and vice versa. Integrating the SDG indicators and monitoring framework as the background to Nexus activities can help to avoid shortcomings from 'cherry picking' of financial investments as well as ensure a more equitable and policy-balanced application of the Nexus approach itself. Conversely, SDGs design promotes multi-sectorial implementation, and the Nexus approach can help make this framework become much more integrated. The key will be to match development cooperation support with country focus that is not on sectors as SDG implementation accelerates.

The following recommendations that have emerged from the expert contribution analysis and discussions are to be taken collectively. As with the WEFE Nexus systemic approach, the promotion of individual technological solutions and mono-sectorial planning need to be integrated and managed as a whole in a Nexus approach. Considered individually, those solutions are not sufficient to successfully implement the whole system of policies, investments, capacities, and techniques that enable maximization of synergies and minimization of trade-offs in land use and spatial planning. Final recommendations are here organized around three overarching questions, two more related to science-technology and one more related to policy:

(1) Where is the knowledge gap and how can modelling help?
(2) What is a Nexus-friendly technology as opposed to classic water management?
(3) How are decisions made that are part of Nexus governance?

5.3.1 Technical/science area: filling the knowledge gap

- WEFE sectors play a fundamental role in determining societal health and economic well-being. However, current and expected changes in climate, growth population, and land-use/cover changes place these sectors under considerable stress. To cohesively improve policies that target these challenges, there is a need for improved integration of economic decision making with biophysical models (Kling *et al.*, 2017 – ref. annex 1). This need reflects an increasing demand to better determine cause–effect relationships in these systems. In order to best integrate individual sectorial models, it is advisable to test this integration by selecting the smallest scale at which allocation of resources is at odds with their interconnected nature, that is, a Nexus hotspot. Another practical solution to model integration is 'soft-linking' or a 'modular approach'. It is in essence a 'loose coupling' in which the output from one module forms the input for the other two as an exogenous factor to the other systems considered.
- When analysing the WEFE Nexus, it is also key not to focus entirely on the technical side of the resource reallocation problem in order not to lose sight of the environmental and social dimensions of sustainability next to economics. Some less than obvious interconnections that should be part of the integrative models developed to analyse the Nexus lie outside production optimization. The use of virtual water and other environmental footprint methods as a comparative analysis tool in this context can help better understand issues of for example, food waste or transport in terms of carbon emissions or biodiversity loss among the environmental impacts along the supply chain and life cycle of a product or activity. There is value in a sectorial model output being considered

for its Nexus implications in terms of social equity as well. Qualitative approaches to modelling can be used in that respect to map the system with multiple sectors and potential beneficiaries and locate priorities and trade-offs.

- Convergence thinking is one such qualitative approach to problem solving that transcends disciplines and integrates knowledge. Adoption would require creating new collaborations among different experts and organizations. Effective collaborations are at the centre of WEFE Nexus implementation. Water managers are being asked to be experts in other sectors to act as bridges between specialists and they will have to be if they want to continue leading the transition to Nexus practice. Silos breaking then starts with education and academic programmes. In order to support new collaborations, the role and availability of data also remains key. Significant economic value can be unlocked by applying advanced analytics to both open and proprietary knowledge. Blended with proprietary data sets, open source can propel innovation and help organizations replace traditional and intuitive decision-making approaches with multi-sectorial, data-driven ones. Creative solutions are needed to complex Nexus problems and a combination of market-driven efficiency and governed planning principles seems desirable.
- The other key integration for WEFE Nexus implementation after sectors is of scales. The scalar perspective is best suited to understanding how a particular system may differ to that for managing it. The national scale may be appropriate for target-setting. The landscape or city scale is important for uniting different needs. But the ideal scale of planning and management for the Nexus is where most economic, sectorial interests congregate (i.e., the basin or sub-basin scale).

5.3.2 Identifying Nexus-friendly technologies

- Sustainable and inclusive intensification and decoupling of resource use and environmental degradation from development can be achieved through technological innovation, recycling and reducing wastage. Technological innovation include desalination based on renewable energy where water is scarce and photovoltaic water pumps, where electrification has not occurred (Hoff, 2011, ref. annex 1). Battery storage systems and/or hybrid technologies such as solar, wind, and fossil fuel combinations can offer flexibility in meeting these water and energy demand of agriculture at times of unavailable supply. There are still technological challenges associated with providing a steady supply but also storage for intermittent use for off-grid solutions. Effective off-grid systems will also require institutional, economic and, to some degree, infrastructural decentralization. Business models and market penetration strategies for these technologies should be explored in more detail to look at potential opportunities beyond cost considerations.
- In multi-use systems, wastes, residues and by-products can be turned into a resource for other products and services. Wastewater-energy integration at treatment facilities and wastewater reuse in agriculture are an example of recycling of water, nutrients, and other resources. Making productive use of the nutrient, organic matter, water and energy content of human excreta and wastewater is also known as productive sanitation. The quality of supplied treated wastewater, however, has to reflect the possible uses for example, for agriculture, domestic or industrial consumption. Existing hydraulic infrastructures equipped with photovoltaic panels provide opportunity for additional low-carbon energy generation with minimal or no extra social and environmental costs.

5.3.3 Policy/development cooperation area

- Evaluating the costs and benefits of selected Nexus solutions according to the priorities of the countries involved in a regional or basin Nexus assessment is seen as desirable. Each region with its own specific context, policy reforms need not create further tensions when the Nexus has been recognized as a risk-sharing element.
- In Central Asia, the notion of the Nexus has seen an evolution in the role played by water in the past three decades. In most Arab countries, Nexus management does not require major institutional restructuring, but rather appropriate changes to protocols. In Latin America, there is agreement that the Nexus approach cannot be reduced to technical and efficiency aspects or decoupled from social inequity. In West and Southern Africa, Nexus support is being injected into reviews of River Basin Organizations' investment and strategic plans and development of operational frameworks for Regional Economic Communities' proposal processes.
- Existing models could be adapted to also assess the impacts of different Nexus investments and contribute to the definition of bankable Nexus projects. A Nexus programme of projects including socially oriented components can become a viable public–private partnership if implemented under a blended finance approach including investment finance from commercial funders and private investors, development capital from DFIs, and grants from donors. Robust risk management system in these Nexus programmes can adequately integrate the complexity of water, energy, and food projects. These partnerships developed around a Nexus portfolio are important to reconcile public and private interests.
- In the short- to medium-term, urban area development represents a challenge because of the increasing concentration of population with significant demands on water, food and energy supply. Cities are where local and global resource constraints meet, passing through the river basin for their most pressing needs. Considering the simultaneous pressures that this type of development will have at all scales, the Nexus approach represents a natural framework for key stakeholders to negotiate sustainable scenarios that minimize trade-offs among the different sectors. Joint ministries for example, of energy and water, and in some instances mining and environment as well, and inter-sectorial consultations for large infrastructure projects are a token of the change in the degree, with which the Nexus approach has been taken on board and operationalized at the national level. The way these ministries are compartmentalized in the respective departments, however, still speaks to the implementation challenge.
- Effective collaboration to deal with the Nexus is deemed more important to achieve than the establishment of an ad-hoc Nexus body. This can be triggered by key issues of the day such as climate change, which have led to the formation of soft-type bodies such as national committees. There is also more value for overarching frameworks such as regional economic commissions, river basin organizations or inter-ministerial committees in a transboundary setting compared to the national context. However, these transboundary organizations may have to undergo institutional fit.
- The first step in the process of identification and implementation of Nexus projects is the analysis of the benefits of a Nexus approach on the different individual sectors. What is not a Nexus project, for example, is when there is a clear predominance of one sector in making the decisions or where there are limited cascading effects from these decisions, i.e. the project is not critical for other sectors. The same applies to the definition of Nexus-friendly technologies. Where these or equivalent frameworks exist in countries that promote inter-sectorial processes of collaboration

from design to decommissioning of engineered and social infrastructure, Nexus projects developed according to these criteria may be rewarded with finance, including international.

In the context of the *United Nations Framework Convention on Climate Change* UNFCCC, there is a potential to leverage National Adaptation Plans as a process through which developing countries can implement or identify their 'nationally determined' adaptation priorities, and in turn how countries might leverage these commitments to adaptation in support of achieving the SDGs. In turn, the implementation of the SDGs, and SDG reporting by UN Member States, must be seen as an opportunity to promote a Nexus approach and vice versa. Effectively a subcomponent or a cross-cutting principle of the 2030 Agenda, the Nexus is perhaps a reminder of the operational needs dictated by action under SDG 2, SDG 5, SDG 6 and beyond. It is undoubted that the water sector can lead the way on Nexus implementation but will not succeed without other sectors sitting at the table.

Scaling up the Nexus approach is advisable as it has many aspects that will lead to more socially, economically, and ecologically sustainable projects. The Nexus can also be a strong support for reaching the SDGs. As for any innovation that needs to be upscaled, it is important to consider a few questions in this context:

- **Identifying the good examples** that have the potential for upscaling. While there are many examples of nexus approaches in the literature, there is a need for a structured approach to further identification and documentation of nexus examples/case studies.
- **Scalable**: Are all elements of a given Nexus programme scalable? Programmes are always implemented in a specific economic, social, and physical context that will influence the implementation and may be different in the next programme. A Lessons Learned report can be used to identify the successful Nexus elements that can be up-scaled.
- **Vertical upscaling**: This deals with the institutionalization of the Nexus approach at different levels. This involves political will and capacity building at the relevant levels and establishing or uses semi-permanent collaborative/governance structures for relevant stakeholders. This requires that the relevant stakeholders are willing to participate in upscaling, and that there are some real Nexus programmes that such collaborative forums can implement. Nexus experts emphasized the value of building Nexus dialogues around existing governance frameworks rather than creating any specific Nexus forums.
- **Horizontal upscaling**: This is about replication of the Nexus programme in another geographic or thematic context. In recent years, Nexus pilot programmes may have been implemented using external resources, both financial and human, but for an actual upscaling to take place, a greater degree of own funding and own human resources are needed. Peer-to-peer exchanges of successful examples and lessons learned are certainly helpful in horizontal upscaling.

Timing: If upscaling is expected to be based on an ongoing or planned Nexus programme and impact on the 2030 Agenda for Sustainable Development, it would be very useful to plan for this as early as possible and align it to the relevant planning processes. Planning cycles in government structures at all levels and in the private sector need to be followed, and such cycles can easily be up to 18 months. Having upscaling in mind right from the start of a Nexus programme will also facilitate the monitoring and assessment of up-scalable programme elements.

Appendix I

List of relevant policy documents

There is a wide body of development policy and practise documentation, academic literature, guidance material on the Nexus approaches. This appendix lists few reference documents considered relevant by the editors.

Reference 1	The New European Consensus on Development 'Our World, Our Dignity, Our Future'
Description	Joint Statement 7 June 2017 by the Council and the Representatives of the Governments of the Member States meeting within the Council, the European Parliament and the European Commission. The key document on EU development policy.
Reference 2	Hoff, H. (2011). Understanding the Nexus. Background Paper for the Bonn 2011 Conference: The Water, Energy and Food Security Nexus. Stockholm Environment Institute, Stockholm.
Description	This paper for the Bonn 2011 Conference presents initial evidence for how a Nexus approach can enhance water, energy and food security by increasing efficiency, reducing trade-offs, building synergies and improving governance across sectors. It also underpins policy recommendations, which are detailed in a separate paper.
Reference 3	Kling C.L., Arritt R.W., Calhoun G. and Keiser D.A., 2017. Integrated Assessment Models of the Food, Energy, and Water Nexus: A Review and an Outline of Research Needs. Annual Review of Resource Economics, Vol. 9:143–163 (Volume publication date October 2017). First published as a Review in Advance on April 3, 2017 https://doi.org/10.1146/annurev-resource-100516-033533
Description	The review highlights the need for integrating biophysical and economic models of the FEW (Food, Energy and Water) Nexus, as well as improved model validation.
Reference 4	FAO (2014): The Water-Energy-Food Nexus. A new approach in support of food security and sustainable agriculture

Description	The FAO paper detailing how FAO sees the Nexus concept in the light of food security in a global context. Introduction to the Nexus concept.
Reference 5	Kougias I., Szabó S., Scarlat N., Monforti F., Banja M., Bódis K., Moner-Girona M., Water-Energy-Food Nexus Interactions Assessment: Renewable energy sources to support water access and quality in West Africa, Luxembourg, European Commission, 2018, EUR 29196 EN, ISBN 978-92-79-84034-0, doi: 10.2760/1796.
Description	This JRC Technical Report examines the potential synergistic benefits to energy, water and agricultural production practices in Africa, arising from an appropriate use of clean energy sources.
Reference 6	FAO: Walking the Nexus Talk: Assessing the Water-Energy-Food Nexus in the Context of the Sustainable Energy for All Initiative July 2014
Description	This report proposes a way to carry out a water-energy-food Nexus assessment approach in order to: (a) understand the interactions between water, energy, and food systems in a given context, and (b) evaluate the performance of a technical or policy intervention in this given context. The ultimate goal of the WEF Nexus assessment is to inform Nexus-related responses in terms of strategies, policy measures, planning and institutional set-up or interventions.
Reference 7	Sustainable Energy Handbook – Module 2.4 – Water-Energy-Food Nexus
Description	This handbook provides a brief overview of DEVCO activities on the water-energy-food Nexus and how this approach will be increasingly used and implemented across a number of thematic units. It is also designed to be a useful summary on the concept itself, with explanation on how it will be operationalized at a global level not only by DEVCO but in conjunction by other EU Member States, international organizations and NGOs.
Reference 8	A Nexus Approach for The SDGs – Interlinkages between the goals and targets
Description	A presentation on interlinkages between and among SDGs – part of SDG TOOLKIT, an initiative co-funded by EU to engage European NGOs at National and European level on the Sustainable Development Goals.
Reference 9	Nexus Message on Water-Energy-Food-Climate through an Urban Lens – Building Integrated Approaches into Implementing the Sustainable Development Goals
Description	Outcome Message from the 2018 Nexus Conference, Water, Food, Energy and Climate, Water Institute at the University of North Carolina at Chapel Hill, North Carolina, USA, April 2018.
Reference 10	Introduction to the Water-Energy-Food Security Nexus. 5 June 2018
Description	Training Module 1 'Introduction to the Water-Energy-Food Security Nexus' provides a theoretical introduction to the concept of the WEF security Nexus. The module has been developed by the GIZ Nexus Regional Dialogues Programme in cooperation with the Institute for Technology and Resources Management in the Tropics and Subtropics (ITT) of the Cologne University of Applied Sciences. Co-funded by the EU.
Reference 11	Messages from the Bonn 2011 Conference: The Water, Energy and Food Security Nexus – Solutions for a Green Economy
Description	Summary document setting out the key messages from this landmark conference on Nexus opportunities, principles, and how to make it work.

Reference 12	UNECE, 2015. Reconciling resource uses in transboundary basins: assessment of the water-food-energy-ecosystems Nexus.
Description	This report from UNECE describes the application of the Nexus approach and in particular how to perform the Nexus assessment in a transboundary river basin.
Reference 13	UNECE, 2017. Deployment of Renewable Energy: The Water-Energy-Food-Ecosystem Nexus Approach to Support the Sustainable Development Goals
Description	This report focuses on the interaction Nexus but with regards to renewable energies. It provides good practices from three cases studies in south-eastern Europe and central Asia.
Reference 14	Udias A., Pastori M., Dondeynaz C., Carmona-Moreno C., Ali A., Cattaneo L., Cano J. A decision support tool to enhance agricultural growth in the Mékrou river basin (West Africa), JRC Technical Report (JRC110346), Ispra (VA), Italy
Description	The paper describes an operational decision support system to help local managers assessing the WEFE Nexus. The e-Nexus has been applied in the transboundary MEKROU River Basin shared by Benin, Burkina Faso, and Niger. The e-Nexus has accompanied policy makers as part of the MEKROU's Strategic Development Plan by cross-checking development scenarios.
Reference 15	**Barchiesi, S., Carmona-Moreno, C., Dondeynaz, C., Biedler, M. (Eds.), 2018. Proceedings of the Workshop on Water-Energy-Food-Ecosystems (WEFE) Nexus and Sustainable Development Goals (SDGs),** JRC Conference and Workshop reports, Brussels (Belgium), January 25–26, 2018 (JRC109346), Ispra (VA), Italy
Description	This report reflects the discussion held at the Conference held in Brussels on 25–26 Jan 2018 on the WEFE Nexus concept and its operationalization. The report outlines the importance and benefits of the WEFE Nexus as an approach and methodology for the implementation of the EU development policies and cooperation, integrating governance and management across water, energy, and food security areas, while attempting to balance different uses of ecosystem resources and services.
Reference 16	**Carmona-Moreno C., Dondeynaz C. and Biedler M. (Edts.), 2019. Position paper on Water, Energy, Food and Ecosystems (WEFE) Nexus and sustainable development goals (SDGs),** JRC Technical Report (JRC114177), doi: 10.2760/5295, Ispra (VA), Italy
Description	This report presents studies, with geographic scopes ranging from national to regional and transboundary levels, providing valuable examples to help the EU and its partners to make informed decisions when operationalizing the WEFE Nexus.
Reference 17	**Crestaz, E. Farinosi, F. Marcos Garcia, P. Biedler, M. Carmona-Moreno, C. (Edts.), 2021. The African Networks of Centres of Excellence on Water Sciences Phase II (ACE WATER 2) Scientific activities outcomes.** JRC Technical report (JRC124069), Ispra (VA), Italy
Description	This report summarizes the key ACEWATER2 project scientific achievements of the activities implemented by the African CoEs, supporting institutions, leading experts and the JRC. The focus is on the WEFE Nexus assessment over selected river basins, complemented with the analysis of cross-cutting topics of continental or regional relevance, in good agreement with the priorities identified at AU, AMCOW, RECs, and RBOs level

Appendix II

List of Nexus experts

This publication is the result of a collaborative work that began in January 2018 during a meeting of international Nexus experts in Brussels and was then complemented with lessons learned emerging from various WEFE Nexus projects.

The editors want to thank all the experts, those who participated since the beginning and those who joined us along the way. They understand that their work is not just about synthesizing information, but also about making scientific and technical knowledge available to contribute to more sustainable development and, ultimately, to help create a better world.

Experts	Affiliation
ADAMOVIC M.	European Commission, JRC
AL-ZUBARI W.K.	Water Resources Management Program, Department of Environmental and Natural Resources, College of Graduate Studies, Arabian Gulf University
AMANI A.	UNESCO-IHP
AMEZTOY-ARAMENDI I.	European Commission, JRC
BACIGALUPI C.	European Commission, DEVCO
BARCHIESI S.	European Commission, JRC
BASHEER M.	University of Khartoum, Khartoum, Sudan
BATTISTELLI A.	Independent Expert
BIEDLER M.	UNESCO-IHP

(*Continued*)

Experts	Affiliation
BODIS K.	European Commission, JRC
BOURAOUI F.	European Commission, JRC
CARMONA-GUTIERREZ A.	IE University of Segovia, Segovia, Spain
CARMONA-MORENO C.	European Commission, JRC
CATTANEO L.	European Commission, JRC
CAUCCI S.	Institute for Integrated Management of Material Fluxes and of Resources (UNU-FLORES)
CIMMARRUSTI Y.	European Commission, JRC
CORDANO E.	European Commission, JRC
CRESTAZ E.	European Commission, JRC
DALTON J.	IUCN (International Union for Conservation of Nature)
DIRWAI T.L.	University of KwaZulu-Natal, Pietermaritzburg, South Africa
DONDEYNAZ C.	European Commission, JRC
DUDU H.	European Commission, JRC
DUPONT C.	IHE Delft Institute for Water Education
EL KHARRAZ J.	Middle East Desalination Research Center (MEDRC)
EMBID A.	University of Zaragoza
FARAJALLA N.	American University of Beirut
FARINOSI F.	European Commission, JRC
FERNANDEZ BLANCO R.	European Commission, JRC
FERRARI E.	European Commission, JRC
FERRINI L.	GIZ, Nexus Regional Dialogue in the Niger Basin
FILALI-MEKNASSI Y.	UNESCO-IHP
FRANCA M.J.	IHE Delft Institute for Water Education
GAMAL M.A.	University of Khartoum, Khartoum, Sudan
GHAFFOUR N.	King Abdullah University of Science and Technology (KAUST), Water Desalination and Reuse Center (WDRC), Biological and Environmental Science and Engineering Division (BESE)
GIRARDI V.	European Commission, DEVCO
GRIZZETTI B.	European Commission, JRC
HANNAH, C.	Nicholas School of the Environment, Duke University
HIDALGO GONZALEZ I.	European Commission, JRC
HOUMØLLER O.	PEMConsult
JAEGER-WALDAU A.	European Commission, JRC

(*Continued*)

Experts	Affiliation
JEMBERE K.	GWP Southern Africa, Pretoria, South Africa
B. JIMENEZ CISNEROS	UNESCO-IHP
KABEYA P. K.	Southern African Development Community (SADC) Secretariat
KAVVADIAS K.	European Commission, JRC
KOUGIAS I.	European Commission, JRC
KOUNDOUNO J.	European Commission, JRC
LAAMRANI H.	League of Arab States
LEMESSA TESGERA S.	IHE Delft Institute for Water Education
LIEBAERTS A.	European Commission, DEVCO Expert
LIPPONEN A.	United Nations Economic Commission for Europe (UNECE)
LORENTZEN J.	PEMConsult
MAKARIGAKIS A.	UNESCO-IHP
MARCOS-GARCIA P.	European Commission, JRC
MARENCE M.	IHE Delft Institute for Water Education
MARTIN L.	Facultad Regional Mendoza/CONICET, Universidad Tecnológica Nacional
MICHALENA, E.	University of the Sunshine Coast, Australia
MISHRA A.	UNESCO-IHP
MOHTAR R.H.	American University of Beirut
MONER GERONA M.	European Commission, JRC
MORENO-ABAT M.	European Commission, ENV
MPAKAMA Z.	Stockholm International Water Institute (SIWI), Africa Regional Centre
NTLAMELLE M.	Southern African Development Community (SADC) Secretariat
PASTORI M.	European Commission, JRC
PISTOCCHI A.	European Commission, JRC
SAMIKWA D.	Southern African Development Community (SADC) Secretariat
SARTORI M.	European Commission, JRC
SCHMEIER S.	GIZ, Global Nexus Secretariat of the Nexus Regional Dialogues Programme
SCHMIDT-VOGT, D.	Freiburg University, Faculty of Environment and Natural Resources, Chair of Silviculture
SEHRING, J.	IHE Delft Institute for Water Education, Integrated Water Systems and Governance Department
SENZANJE A.	University of KwaZulu-Natal, Pietermaritzburg, South Africa
SMAKHTIN V.	United Nations University – Institute for Water, Environment and Health (UNU-INWEH)

(*Continued*)

Experts	Affiliation
SZABO S.	European Commission, JRC
TAKAWIRA A.	GWP Southern Africa, Pretoria, South Africa
THIEM M.	GIZ, Global Nexus Secretariat of the Nexus Regional Dialogues Programme
TIRUNEH J.K.	GWP Southern Africa, Pretoria, South Africa
TSANI S.	Athens University of Economics and Business
UMLAUF G.	European Commission, JRC
VAN HULLEBUSCH E.d.	IHE Delft Institute for Water Education
VERBIST K.	UNESCO-IHP
XENARIOS S.	Nazarbayev University, Graduate School of Public Policy, Astana, Kazakhstan

IWA Publishing's authorised EU representative for General Product Safety Regulations is Diane D'Arras, 15 rue Duret, 75116 Paris, France, e-mail: safety@iwap.co.uk.

Printed and bound by CPI Group (UK) Ltd, Croydon, CR0 4YY
12/05/2026
02108869-0008